KEEPING THE PRAYING MANTIS

MANTODEAN CAPTIVE BIOLOGY, REPRODUCTION, AND HUSBANDRY

Orin A. McMonigle

COACHWHIP PUBLICATIONS
GREENVILLE, OHIO

ACKNOWLEDGEMENTS

I would like to thank Christian Schwarz, Richard Trone, Tammy Wolfe, Andrew Lasebny, and Michael McNichols for contributions to development or editing of the text and Rebecca Salutric for help with some photographs. Most of all I thank God for creating the spectacular praying mantis and the opportunity to learn and write about his handiwork.

DEDICATION

This book is dedicated to my loving wife Sylvia and my daughters Kree and Gwynevere.

Keeping the Praying Mantis
© 2013 Orin McMonigle

ISBN 1-61646-166-7
ISBN-13 978-1-61646-166-9

CoachwhipBooks.com

Front cover: Female *Stagmomantis limbata,* from Arizona, with ootheca.
Back cover (top left, clockwise): *Deroplatys lobata* nymph in threat display; *Hymenopus coronatus* subadult (© Tammy Wolfe); and *Brunneria borealis* showing captive coloration, usually found on grass, not bushes.
Unless otherwise noted, all images © Orin McMonigle.
Images with (CC) are Creative Commons licensed, with attribution.

All Rights Reserved. No part of this publication may be reproduced, stored in a retrieval system or transmitted in any form or by any means—electronic, mechanical, photocopy, recording or any other—except for brief quotations in printed reviews, without the prior permission of the author or publisher.

CONTENTS

Introduction	5
Man and the Praying Mantis	14
Mantids and Their Relatives	24
Mantis Biology	34
Mantis Behavior	50
Habitat Considerations	60
Health Issues	69
Mating	76
Oothecae	89
The Preying Mantis	101
USA Mantids	122
Exotic Mantids	144
Glossary	191
Bibliography	195

Ciulfina species from Queensland, Australia. (CC) Donald Hobern

Gonatista grisea, immature, from Florida. (CC) Geoff Gallice

INTRODUCTION

When I began keeping insects, mantids were a first choice as pets because there were beautiful species resembling flowers and leaves and any species from around the globe could be legally imported. As luck would have it, locating livestock was nearly impossible. I would call long distance to a few vendors every month and drive to distant reptile shows in hopes of acquiring one or two expensive specimens of one new species in a year or two. Ironically, today when imports are no longer allowed and exotics are of dubious legal status, captive-bred specimens of dozens of species can be acquired in days with minimal effort and cost. On the wholly positive side all native and adventive species are legally kept and can be shipped across state lines (at the time of writing) though there may be specific regulation in some states or cities (but unlikely since they are not venomous). Expanded interest in mantids as pets has made it easier to acquire standard natives like *Stagmomantis* and *Tenodera* as well as the exotic looking *Brunneria*, *Gonatista*, and *Phyllovates*. European enthusiasts have access to a wide variety of species though some of their natives are protected. (*Mantis religiosa*, the most widespread and common mantis on the planet, is protected in some European countries.)

Wild-caught mantids used to be imported with reptile shipments, but the mantids would seldom survive the extended lapses in feeding and humidity extremes. Dripping wet paper towels were often placed in the deli-cup, so the enthusiast would be lucky to find a female capable of laying an ootheca or two before it fell over dead. Since wild adult females in most cases already mated, it was not too difficult if just one adult female could be found. The most common imports in the 1990s and early 2000s came from Tanzania and included *Parasphendale*, *Pseudocreobotra*, and the occasional *Otomantis* or *Tarachodes*. Rarely a vendor would import a few *Deroplatys* and *Hymenopus* from Malaysia, one of the smaller *Hierodula* from Indonesia, or *Alalomantis* and *Sphodromantis* from Cameroon. Most of these creatures arrived in terrible condition but captive breeding proved easy if just a few healthy specimens could be acquired. Captive-hatched *Acanthops*, *Creobroter*, *Gongylus*, *Heterochaeta*, *Phyllocrania*, and *Popa* nymphs were encountered from the very rare EU import in the early to late 1990s. In the late 1990s to mid-2000s, large *Sphodromantis lineola* nymphs of unknown origin could be found uncommonly at pet shops. It was even more difficult to find any of our interesting native species. The adventive Chinese and European mantids were easy to find but never out of season. The market was tiny and prices low but I made a go of breeding the available stocks and was so

Four-year-old from *Keeping Aliens* with *Sphodromantis lineola* female.

A few years later, a 16-year-old with *Stagmomantis limbata*.

successful I decided to wholesale specimens and was the only established breeder/wholesaler of captive-produced mantids in the U.S. for nearly a decade. There was not much money in it but it was a lot of fun.

In 2001, a friend and I completed a small book on keeping praying mantids that we hoped would spark interest in keeping mantids as pets. We shared our breeding experiences so every hobbyist starting out would not need to reinvent the wheel. In 2004-2005 I began to hear rumors about new regulations coming along resulting from a bill that allowed predatory creatures including non-native mantids to be regulated as secondary plant pests (they could eat a bug that could eat a bug that could eat a plant or feed on a pollinator) signed by Clinton in 2000 and new regulatory interpretations and enforcement paid for by Bush antiterrorism money in 2006. Through 2006 the USDA website said it was perfectly fine to ship exotic mantids interstate since they are considered beneficial insects. Because mantis breeding was for love, not money, I decided to drastically reduce my collection in 2004-2005 and switch over to our more interesting native species because they were not included in the interpretation (anything from tarantulas to chameleons could be included at any time). Increasing interest in this group of creatures has led to regular availability of less common native species and the opportunity to acquire and work with our spectacular *Phyllovates chlorophaea* in 2007. As of printing (2013) I continue to research and document the captive biology and husbandry requirements of U.S.-native Mantodea and continue to search out sources for difficult-to-find natives like *Mantoida maya*. I feel it is important to foster an interest in the study of these incredible

Creobroter sp., male; popular and somewhat common since the early 1990s.

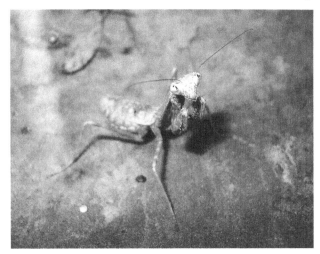
Stagmomantis carolina showing "personality."

Phyllovates chlorophaea nymph.

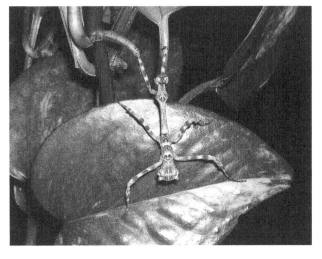
Pseudovates arizonae, our most exotic native.

creatures, and that study should not be relegated to an elite few, so I continue to help moderate the oldest and largest mantis forum and learn things from others. The number of species showing up in recent years that I have never heard of anyone keeping before is amazing. The contemporary selection of species in Europe is unbelievable though they are primarily variations on the same themes (such as a dozen *Sphodromantis* species, where once we had two).

This book was written because there still are no large-format, English-language books for mantids available to the mantis enthusiast. There are two main books available today in the U.S. The big one is *Praying Mantids* (1999) by Prete et al., which is a monster, 363-page, hardcover text on mantis research that reads like a phone book, was not written for hobbyists, has only a few pages relating to husbandry, and a dozen color plates with photos that look like they were taken in the 1970s (and most actually were taken from Edmund's 1972-1976 works). The other is *Praying Mantids: Keeping Aliens* (2001) by myself and Lasebny, which is only 44 pages and mostly black and white.

Late instar nymph, *Liturgusa* sp., from Panama. (CC) Bgv23 / Flickr.com

Juvenile *Sibylla* sp. from Zimbabwe. (CC) Peter Pawlowski

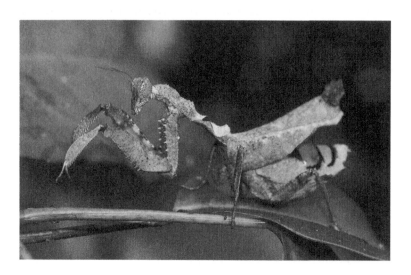

Adult female *Acanthops* sp. from Ecuador. (CC) Geoff Gallice

The text was rather small, to keep down on the page count due to constraints of its construction. It was updated with two color plates, a few text pages, some diagrams and additional ootheca images in 2008, but there was simply no room in the format to attempt a serious overhaul or expansion, so it is essentially the same book from 2001. I flipped the name order since I wrote most of the original text (and more so with the update) but mostly because, though Andy had previously worked actively with mantids and published articles in the Mantis Study Group, he disappeared from the hobby around 2002. *Keeping Aliens* contained cutting edge information that could be found by hobbyists nowhere else when it was written, but much has become common knowledge, some details have changed, and very little has survived in this text.

Other non-periodical literature is either out-of-print and difficult to acquire, extremely short, not in English, or was not intended to provide husbandry information. There have been some small format guides from the UK that are hard to find: one is *An Introduction to Praying Mantids* (1997) by Bragg which is 16 pages long, out-of-print, and even the cover is black and white. The other is *Your First Praying Mantis* (1999) by Willis that is 36 pages only because both sides of the front and back cover are included in the count. The cover drawing makes it look like a children's book, but it has color photos of a dozen mantids still popular today and contains some adequate introductory husbandry information. When I first acquired a copy in 2006 I imagined it might have been a popular guide if the cover were not so ugly and misleading. It was reprinted in 2003 using a *Rhombodera* photograph from the introduction as the cover and the title was changed to *How to Care for Your Praying Mantis*, but that version is even more obscure. The rarity of this book may have resulted from no references to the author in the hobby and it may have been written secondhand. A small text I only learned about recently called *Rearing and Studying Praying Mantids* (1980) by Heath and published by the UK's Amateur Entomologists' Society was revised in 1990 for a total of 22 pages with nine black and white photos. I have not seen either version for sale. Foreign husbandry texts include the small-format, 191 page, German *Praxis-Ratgeber. Mantiden – Faszinierende Lauerjäger* (2001) by Bischoff et al., and the larger-format 70 page, Danish *Knaelere* (2007) by Larsen. Other related publications include two catalogs (Ehrmann 2002, Otte & Spearman 2005, the first one in German also features 194 color photos), an English-language field guide to *Mantids of the Euro-Mediterranean Area* by Battiston et al. (2010), and a 521 page German-language monography on *Mantis religiosa* by Berg et al. (2011). Lastly, there are some different children's books that probably contain no more than a few hundred words combined, but have likely sold more copies than all the above. The internet presently contains volumes of information, but it is a different media, can require excessive time to locate specific information and sort the good from the bad, and it can not be enjoyed or held in the hand in the same way. Thanks to Coachwhip Publications, *Keeping the Praying Mantis* is the only large format, English-language, full-color book on keeping or rearing praying mantids ever published. This is the realization of what I had hoped for in *Keeping Aliens*.

The mantis hobby remains small (even in Europe) so the normal cyclical interest in species combined with the great effort it takes to keep an adequately large colony of a single species in perpetuity have led to the loss of many species. Of course this holds true for our

Deroplatys desiccata subadult female.

natives here in the USA, where *Pseudovates arizonae* has never stuck around because almost nobody can get even the F1 generation to mate successfully. I sent out hundreds in the mid-1990s and early 2000s and they have been introduced and extinguished in the hobby at least three other times. The large *Stagmomantis floridensis* showed up a few years ago and was probably as easy as other *Stagmomantis* but was lost from culture before I had a chance to acquire any. When I started keeping mantids I knew if I chose to get rid of a species or accidentally lost it, that stock would never be available to me again. My most memorable loss came after sending out thousands of *Acanthops* over about a five-year period. I had made the mistake of selling off too many combined with feeding a bunch of males to a really nasty female. I searched around and could not find them at any price. Today, things are much better because it is often possible to locate a mate from a fellow hobbyist, but stocks are still lost due to cyclical interest. *Phyllovates* stock is barely hanging on after six years, while *Gonatista* has rarely been available past F1.

Our handsome native *Thesprotia graminis* and *Oligonicella scudderi* are easy to keep and breed but nobody (including myself) has bothered to keep them past a generation or two.

The longer a person keeps mantids, even the same stock of the same species, the more it seems there is an exception to every rule and time frame. Nevertheless, compared to some insects (such as rhinoceros beetles), mantis development, gestation, and reproduction are like clockwork. The suggestions and rules of thumb regarding habitats and prey outlined here will allow for success but are not the only successful methodologies ever discovered. (There may be more than one way to decapitate a mantis, but the vast majority do not result in successful reproduction.) I have attempted to qualify all statements and offer conflicting reports, even those that are obviously inaccurate, however common. I have tried to avoid over-qualification, which can make it seem like there are no real answers, when there are. I have also tried to make the main husbandry sections long enough to include all important information, but not so long they could not be read through in a sitting.

In this text, mantis is used for singular and mantids plural since it is common usage today and for attempted consistency (Helfer 1963, 1972; Lasebny & McMonigle 2001). The original terms were mantis and mantises but in the mid-1900s someone decided to add the terms mantid and mantids because they were all Mantidae in the Orthoptera. Some recent texts misinterpreted the original usage and tried to eliminate mantis as the common name, possibly so as not to confuse the common term with the genus *Mantis*. Of course the general public and spell checkers never got the memo so most people have only heard the term mantis. (All four terms are used in the Prete et al. 1999 book

KEEPING THE PRAYING MANTIS

Hymenopus coronatus. (CC-share-alike) Luc Viatour

Sibylla pretiosa. (CC-share-alike) Luc Viatour

Empusa fasciata. (CC-share-alike) Alastair Rae

West Papua mantis. (CC-share-alike) Guido Bohne

Ameles spallanzania, European dwarf mantis. (CC-share-alike) Ferran Pestaña

Sphodromantis lineola silhouette. (CC-share-alike) Neil Parker

depending on the author of the chapter, but in combination in older texts as far back as Stefferund 1952.) Attempts at clarification by subtle changes to established names result only in muddying the waters and are influenced by the constantly evolving ideas regarding the hierarchy of this group of predatory insects. Even the unrelated mantisflies were changed to mantidflies, though the family spelling is still Mantispidae rather than "Mantidpidae" as consistency was never a consideration. With standard usage for converting a family name to a common name only, the Mantidae would be mantids, but the elevation of the family to a suborder, then order, meant a number of subfamilies were now to become families. A flower mantis like *Harpagomantis* would still be a mantis but would now become a hymenopodid and could no longer be called a mantid. Technically, no species of empusid, tarachodid, etc., can be called a mantid since it is no longer a representative of the family Mantidae, but this underlines the confusion in trying to merge a technical common name and common usage common name. Confused by the mess, it is not terribly rare to see "stagmomantid" and "sphodromantid" thrown around. Of course then there are the terms dictyopteran, mantodean, and whatever changes are made to taxonomic hierarchy in the future. Calling this text *Keeping the Praying Mantodeans* was only very briefly considered. Not every new name can be blamed on taxonomic fluidity. It is not rare for a writer to confuse the terms and spelling to end up with the name *preying* mantis. My favorite was when I discovered my youngest daughter had for years thought I was saying pretty mantis.

Pseudomantis albofimbriata from Australia. (CC) Donald Hobern

I have been greatly encouraged by and thrilled with the growth of interest in keeping mantids since publishing *Keeping Aliens*. I would like to think the guide played a part. There are a number of dedicated forums and enthusiasts who have chosen to stay in the hobby for many years. Hopefully this book will spur some renewed and new appreciation for these creatures of myth and lore. The vast majority of information presented here is based on first-hand experience keeping the listed species over multiple, consecutive generations. Photographs and information are included for a number of U.S. and exotic mantids. Details for a few more recent arrivals are included here thanks to friends in Denmark, Germany, and the U.K. These chapters contain detailed tips and techniques that will help the beginning enthusiast successfully keep mantids and may even teach the expert a trick or two.

MAN AND THE PRAYING MANTIS

The word mantis comes from the ancient Greek word μάντις, which means prophet or soothsayer. This group of insects was named because the position of the forelegs is suggestive of prayer. The Greeks thought they had supernatural power while Moslems believed mantids could be used as a sort of compass because the mantis' head would always face towards Mecca when the forelegs were held in prayer (Hutchins 1966). Recently, an author (in Prete et al. 1999) claimed ancient people would never have chosen the name had they known mantids were predatory. This seems to be a mistake based on modern Western assumptions. I think it is amazing to imagine ancient people could be ignorant of the mantis diet. These peoples did not have television nor video games and did not spend the summer indoors in air conditioning where they could be wholly ignorant of the activities of the creatures that surrounded them. Also, the western notion that food's primary curse is its overabundance would have been lost on people who prayed for fear of starvation. Ancient prayers centered around feasting and food more than they do today so the fact mantids never forget to pray before and after every meal would seem pious indeed.

Human fascination with mantids is evidenced by the countless stories people tell. During insect displays and shows, children and adults always have stories to share about these endearing creatures. The story of the forgotten mantis ootheca which later hatches into a "million" nymphs that run crazy around the house and are found for weeks is told again and again. A child who shared hot-dogs and soda pop with a newfound mantis friend is another common story. The ex-army private telling of his platoon's giant mantis, kept as the pet/mascot for a few weeks during a stay in Asia, is reminiscent of the previous story except that the troops would offer their pet hamburger, and beer poured into a bottle cap.

Mantis head. (CC) Sam Fraser-Smith

Another favorite story is: "When I was younger they were everywhere but today you just don't see them anymore." In some areas of the world where habitat destruction is erasing

"A million" mantis hatchlings.

species from the planet every few seconds this story might be true (though the widespread species commonly encountered by humans are masters of disturbed and grassland habitats increased by human activity, in which case pesticide use would be the only credible threat). However, in the eastern U.S. the story references *Tenodera* and *Mantis* and the culprit is most certainly the amount of time spent outdoors. Not only do few children spend time outdoors like they used to, but also, adults do not seem to realize they used to actually look outside the front door.

The "endangered and protected" myth seems to be the most pervasive story, at least in the eastern United States. I remember a few times on the playground when children would scream to other children, "You can't step on it, it's illegal to kill a praying mantis!" I cannot count the number of times I have been told and asked about the protected status of mantids. I remember that was my first question when I spoke to an entomologist while touring a university collection in my teens. He wrinkled his brow and simply said, "No, of course not." He smiled but for a moment looked as if I had asked if the moon really was made out of cheese. Last I heard there was a temporary law in the early 1900s in one area of New Jersey which protected mantids during an introductory period for crops but the evidence for this story is shaky at best. In Florida, the *Sarasota Herald-Tribune* (March 9, 1972) reported the Florida Senate passed a bill that would make the praying mantis the state insect and protect it, but the bill must not have passed the house. For a time I would explain the protected story is not true and the mantids people see normally are from Europe and China anyway. However, the main result is a reluctance to kill mantids and my protestations are not likely to combat a highly pervasive myth, so I now just say, "Yes, you should never kill a mantis."

Many gardeners and farmers who are wary of increasing arthropod resistance to pesticides and their danger to humans employ mantids as a natural and chemical free method to reduce pest numbers. They were well-known insect helpers long before they were listed as "Insect Friends of Man" alongside lady beetles in the USDA's *Yearbook of Agriculture 1952*. In the U.S., the most commonly encountered mantids were brought over from Europe and Asia for their pest control abilities long ago (or accidentally on nursery stocks, depending on which theory you prefer). *Tenodera sinensis* was introduced about 1896 (Stefferud 1952) and *Mantis religiosa* about 1899 (Swan & Papp 1972), while the less common *Tenodera angustipennis* was first reported from Delaware and Maryland in 1926 (Swan & Papp 1972, Arnett 1993). *Iris oratoria* was also introduced about 1933 (Helfer 1963). Beyond the initial introductions, the Chinese mantis and European mantis were known to be deliberately introduced to areas across the U.S. at least fifty years ago (Helfer 1963) and probably well before (Stefferud 1952). Mantis oothecae have been a popular fare sold by nurseries and garden shops for natural pest control for the last forty years with the advent of popularity in integrated pest control (Frye 1992). Powell & Hogue (1979) state that *T. sinensis* and *I. oratoria* oothecae were the species most widely sold to gardeners for pest control in California. *Iris oratoria* was commonly used for pest control in cotton fields. In Texas, *T. sinensis* has been employed to combat cotton bollworms (Ross 1984). Though usage is limited to certain prey and affected by specific circumstances, a number of different scientific studies listed in Hurd (1999) have shown mantids reduce herbivore populations while Moran et al. (1996) demonstrated a generalist arthropod predator can improve plant productivity even within a complex natural ecosystem.

I have observed mantis pest control prowess firsthand. When I moved into my present home there was a newly created grass bowl field which produced plagues of no less than seven species of grasshopper year after year. Had I been a farmer I would have been greatly saddened, but I was in my glory. I could go outside for a short while and collect all the food I could use (for my charges) and it was a joy to observe all the different species and development. (One species would hatch purple nymphs the last week of February when it was rarely above 40° F.) In late summer of year five, I noticed a few *Tenodera sinensis* in the field eating the grasshoppers. These mantids probably made their way from some other local area (the closest field is on the other side of a forest preserve a quarter mile away) since the species is common here, but an ootheca could have been released in a nearby garden. The following year there were far fewer grasshoppers, but plenty to collect for some feeding purposes, and there were many more mantids. In the subsequent four years only three species of grasshopper have been observed and the population so reduced it takes effort and time to catch a single one, like most older fields in the area. There are other possible explanations in an uncontrolled wild setting but the only creatures I observed that would affect grasshopper populations were soldier beetles (Cantharidae) and wolf spiders (Lycosidae), and those were only common years one to six. (The wolf spiders had been present in plague proportions, but now are seen only when rocks are flipped over.) I would trade in this powerful anecdote to get my grasshoppers back.

"The value of insect predators such as ladybird beetles, mantids, and lacewings is legendary" (Imes 1992). As with all legends there will be detractors. A few naysayers in recent years have tried to claim that the pest control abilities are limited because the number of mantids in a season cannot keep up with the exponential growth of pest populations in a given summer.

Various claims have not been based on multiple or long-term studies but rather on a functional response graph for 1st instar *Tenodera sinensis* with unidentified flies (Hurd 1985) combined with a May 31st to July 18th, off season, single predator, single-habitat data set from the same author (Hurd & Eisenberg 1984). If Hurd had run the same functional response test on adult females with ravenous appetites rather than 1st instars he could never have concluded they are "unlikely to be good control agents for pests that achieve maximum density late in the season." Unfortunately he made no attempt to quantify consumption rates relative to development, though it is hard to imagine he was unaware of the exponential expansion of the *T. sinensis* appetite as it matures. His accompanying data sets (Hurd & Eisenberg 1984) are from part of a single season, old field, five-date count of immature Gryllidae (crickets seem an odd choice since he writes he knew crickets remain on the ground and would be unlikely to be encountered by *T. sinensis*, but yet by sheer coincidence offered the only statistically significant variation between the experimental and control groups) and *Acrididae*, grasshoppers. The conclusion based on just five survey dates on two local sites, charted with prey ranked by size, was *T. sinensis* did not affect the number of grasshoppers (in essence, do not eat them) and cricket populations expand greatly at least on June 12th and 19th but return to normal by June 30th. The last survey of his experiment graph was dated July 18th, long before *T. sinensis* females in the northeastern U.S. could be large enough to be expected to eat grasshoppers in quantity. (The animals would likely not even be subadults yet.) A critical look at the timing of the experiment's end begs the question if the authors were ignorant of the acridid and *T. sinensis* life cycles in the area (though Hurd (1985) says in Delaware *T. sinensis* females barely have time to produce a single ootheca before winter), were unaware that a predator could switch prey types at as it develops, or if the purpose of the experiment was to artificially force a desired conclusion through manipulation of the experiment's time frame. An established field habitat was also chosen as the only habitat type, which also seems calculated. The effects of *T. sinensis* predation would be moderated by greater prey variety and established acridid predators. The experiment may have demonstrated that immature mantids in those field plots did not feed on immature grasshoppers, but the results were extrapolated out to conclude older and mature female mantids would not eat grasshoppers, which was a grievous overreach. The experiment (Hurd & Eisenberg 1984) was criticized by Hairston (1989) for "unrealistic density" and later Hurd and Eisenberg (1990) conducted a field enclosure study in which *T. sinensis* did eat grasshoppers. Later, Hurd (1999) says mantids reduced the number of field hoppers and allies in all studies, not mentioned in Hurd (1985), and "Though their effect on the rest of the arthropod assemblage may be diffuse and sometimes hard to measure, yet they are capable of benefiting the plant community by reducing herbivory." However only the 1985 paper seems to ever be cited.

It is important to know the original experiment was supposedly never meant to measure the value of *T. sinensis* as an agent of pest control, which was the reason it was not performed on crops or gardens. It was designed to study the ecological effects on an old field, so the appearance of bias can be pinned mostly on its misapplication. However, Hurd and Eisenburg's (1984) odd cricket (Gryllidae) results illustrate possible bias in the conclusion. They did not conclude *Let's try another experiment*

Female New Zealand praying mantis, *Orthodera novaezealandiae*. (CC-share-alike) Bryce McQuillan

Deroplatys desiccata.
© Kenneth Tinnesen

Sibylla pretiosa.
© Kenneth Tinnesen

since there may be some external variable we did not account for, but rather, "*T. sinensis* ... actually enhanced cricket abundance in experimental field plots during early to mid-June" (Hurd 1985). In defense, Hurd and Eisenberg (1990) recorded a later experiment in which the cricket population did not increase.

Myths unfortunately tend to generate new background stories as they grow. Other recent negative press about the mantis pest control ability is difficult to address because there was never any evidence for it. In the same paper where he claimed the earlier experiment showed mantids do not eat grasshoppers (Hurd 1985), as an aside he mentioned *T. sinensis* females that just happen to "find themselves" on flowers late in the season could eat bees and wasps. Hurd did not claim to have any data for this statement and pointedly stated the female mantids "just happened" to be on the flowers in order to keep in line with his earlier claim in the paper that *T. sinensis* are unlikely to disperse to areas of high prey density. Hurd (1999) says *T. sinensis* seem to spend no more time on flowers than any other vegetation. A few naysayers in recent years jumped on the pollinator suggestion and use his off-the-cuff comment (which he claimed no data or evidence for) as "scientific evidence" by citing his paper while dismissing the contradictory half of the same sentence—the mantids only "found themselves" on the flowers. Of course there has been no attempt made to quantify differences in pollinator consumption or pollination rates, let alone study differences between single and multiple years. However, this at least can be partly justified since, in most food growing areas, seasonal crops are already growing long before a mantis could be large enough to catch a pollinator, so the expense would be unjustified and an honest pollination experiment would require multiple years' data based on seasonal variation alone. A more recent herring tossed into the same "argument" is beneficial insects may also be taken, but there is no reason to imagine the odd run-in with a lady beetle or lacewing would affect predator populations since prey density exceeds that of predators, and Chrysopidae and Coccinellidae do not occupy the same niche as the large Mantidae species normally used in pest control. (Mantids have no practical application to controlling aphids, scale, whitefly, etc., because they rapidly outgrow small prey, with exception of *Miomantis* employed for aphid control on specific greenhouse crops where standard pesticides and ladybugs were not an option.*) Hurd (1985) concluded, "... I am loathe to extrapolate my observations of *T. sinensis* to mantids in general," but unfortunately this was another of his "lost" statements and his comments have not only been extrapolated far beyond what they did not demonstrate, or showed for a partial season, single habitat type, and single species in its early development stages, but also are frequently extrapolated out to the entire order by various authors citing his paper (but not reading it or subsequent works). Sadly, not all distorted stories about mantids are positive. Some may even be crudely fashioned to support bureaucratic power grabs.

Misinformation is a pervasive aspect of mantis information. Whether they are considered cow killers, cause blindness if they spit in one's face (Hutchins 1966), are attributed uniquely unlikely husbandry requirements by some hobbyists, or unbelievable superpowers

---

* Results reported by two commercial operations in northern California that purchased *M. paykullii* for aphid control 3-4 times a year over a four-year period, 1997-2000. Coccinellidae could not be used because of compatibility issues with the specific crop.

Blepharopsis mendica. © Kenneth Tinnesen

Texas unicorn mantis laying ootheca. © Tammy Wolfe

by various entomology texts (such as headless mating requirements or deimatic displays becoming "deadly flowers"), humans love to make up stories about the praying mantids. One interesting tale in an entomology text (which correctly describes the "devil's-flower" as a cryptic species) claims there is a South American mantis that resembles a stick-insect and it joins groups of the dull-witted vegetarians in order to devour them (Hutchins 1966). Other mantis myths are discussed in later chapters. There certainly is misinformation for other types of invertebrates, but none with stories so spectacular, pervasive, and varied as the mantis.

Science fiction movies have cashed in on man's fascination with the appealing and alien look of the mantis. *Mimic* (1997) starred human-sized mantis hybrids that resembled a bum up until the unwary person was too close to avoid the raptorial forelegs. (You would think a streetwalker would make more sense as bums are not highly attractive.) In *Starship Troopers* (1997*)* the main troops of the alien bugs resemble monstrous swarming mantids, complete with giant slashing front legs. Before such high tech special effects, movies used real mantids superimposed over shots of buildings to make them appear two stories tall (*The Deadly Mantis*, 1957). These movies would certainly have been less popular without the powerful and alien appearance of the mantis.

Even when they are not starring in film, mantids engender the term *alien* in the human mind. Certainly they have been around longer than humans. The reference is to behavior and appearance, which seem far removed from the world of man. A mantis feeding on a struggling cricket is interesting to observe, whereas a big cat chewing apart a panting wildebeest is disturbing and not at all fun to watch for the average person. Mantids can lose large parts of their bodies (or heads) and still temporarily appear to be wholly unaffected. Additionally, people who claim to have seen extraterrestrial aliens often detail creatures with very large eyes and triangular heads. The aliens are often green or gray.

The mantis may be the most popular mascot of general entomology texts. There are many dozens of books that have chosen the mantis as the featured cover insect because of awareness of the draw of the mantis. A few examples: *Grasshoppers and Mantids of the World* (1990), which has more photographs and information about cockroaches than mantids, and *Alien Empire* (1995) have the exact same cover; a mantis photograph of the face and forearms of *Polyspilota aeruginosa*. *The Practical Entomologist* (1992) displays a curious mantis on the cover but the information on mantids is limited to two paragraphs hidden in the back under the title "minor insect orders." *How to Photograph Insects & Spiders* (1994) has a mantis cover and just a single mantis photo among the ninety four images within. When I visited a bookstore recently I noticed a few newer titles with mantis covers. *Insectopedia* (2010) has a mantis on the cover and just a few paragraphs in the book specific to mantids, while *A World of Insects* (2012) has a beautiful *Rhombodera* in threat display on the cover, but I could not find a single mention of mantids in the index. Oddly, despite the large number of books with mantis covers, very little mantis content is usually included.

Representatives of the order Mantodea have a number of characteristics that make them great pets. First and foremost, they respond to the keeper and keep a watchful eye out. Few, if any, other invertebrates seem to interact at such a level. Second, they are mesmerizing to look at and quite beautiful. Even some of the most ordinary species are works of art, painted with various hues and patterns, and uniquely structured. Third, most mantids need relatively little space. Caging requirements are minimal compared to marine invertebrates, large reptiles, cats, dogs, etc. Cages can be made from various containers that can be obtained for free. Fourth, unlike vertebrate pets such as cats, birds, gerbils, etc., that can infect humans with a multitude of diseases, mantids do not spread diseases to humans. Fifth, cages rarely stink and seldom need cleaning. Sixth, they have short natural lives so it is quite difficult to grow bored with them. Seventh, they are a great introductory pet for children to learn about the natural world and responsibility. Eighth, not a single species of mantis exudes poison or injects venom. (The defensive bite or pinch may be startling but causes no real harm and is easily avoided by not agitating them.) Ninth, none have unpleasant defensive odors common to other orthopteroids. Lastly, with exception, they do not have to be cared for and fed every day.

The history of studying the reproductive biology and husbandry requirements (manticulture) may extend back to before recorded time, but only in recent years has there been any sort of organization or documentation. Until recent decades keeping was restricted to maintaining an adult animal for a few weeks or encouraging development in the garden. Helfer's (1963) book on U.S. orthopteroid insects says the Chinese mantis is an interesting pet, while Swan & Papp's U.S. insect textbook (1972) claims, "They make interesting captives (or pets?)." I remember collecting Chinese mantids and keeping them as a child and teenager but my first foray into true manticulture was with *Hierodula patellifera* stock purchased in 1993 and kept for consecutive generations. Within a year or two I found others to correspond or talk with about mantis husbandry and development. Meanwhile, on May 18, 1996, the first meeting of the Mantis Study Group was called in the U.K. (Bragg 1996; this was also the first meeting of the Blattodea Culture Group). The MSG flourished, published numerous articles in its quarterly newsletters, but essentially burned out terminally by early 2001. In 2003-2004 the first two incarnations of Mantidforum.net began and quickly showed the great enthusiasm people of like minds can have. I paid for its upkeep for a few years with the meager earnings from *Keeping Aliens* but it was not enough. In 2007 trouble with software and spambots had long been taking their toll on the forum, myself, and moderator Richard Trone, and was growing worse. I turned to a friend I knew through keeping mantids in the late 1990s, Peter Clausen. He took the forum, fixed the software issues and made improvements, including a yearly mantis calendar contest (with 2013 just completed). Maintaining a community of enthusiasts is very important for future conservation and interest in the fate of this order, as well as to maintain bloodlines of rare or rarely encountered native species. In Germany, the IGM (Interest Group Mantodea) began a few years ago and has tried to keep track of various stocks through a numbering system. Beyond a few *post mortem* twitches of the MSG after 2001, an unrelated, online, U.K. mantis newsletter edited by Gillian Higgins has run nine issues from Aug. 2010 to Dec. 2012. A good number of mantis husbandry articles have run in *Invertebrates-Magazine* (USA) 2001-2012 and *Exotiske Insecter* (Denmark; some translated for a five issue English-language run ca. 2002).

Praying mantids are, without question, the most loved of all invertebrates. Nearly everyone has a childhood story about these strange creatures. The use of mantids for biological pest control is well known and highly praised. Mantids make great pets for children as well

KEEPING THE PRAYING MANTIS

My Mayor Mantis pin collection.

as adults. Humans continue to dedicate things they build to the mantis. There is a martial arts form, minor deities, and a popular garden tool named after these venerated creatures (all using the common name of mantis, not mantid). Oothecae and exuvia are ground up and eaten in hopes of curing a variety of ills in Chinese herbal medicine (Ross 1984). My favorite video game, *Escape from Bug Island*, has a giant mantis on the cover, and among various monsters features killer mantids that ooze from gigantic oothecae hanging in the trees (though the mother is never encountered). Even closer to home, my neck still hurts from the Mantis roller coaster, while the mascot of the largest local bug festival is Mayor Mantis. Books are even written about keeping them as pets. This chapter is not meant to be exhaustive and offers just some examples reflecting man's infatuation with the praying mantis.

Many of us spent years of childhood in fields and creeks hunting for all manner of invertebrate. The monstrous praying mantis was among the most spectacular of finds. It is curious to wonder if today's children will develop an appreciation for our world's spectacular microfauna when informative years are more often spent on electronic devices. Hopefully the artificial world would not hold their attention so long that future children will be surprised to learn the mantis is or was ever a real creature.

MANTIDS AND THEIR RELATIVES

There are related and unrelated insects that look very much like mantids. Stick-like mantids can be difficult to discern from various plant-feeding walkingsticks (Order Phasmida) but the tibiae of the thin front legs are able to fold down and are armed with raptorial spines to catch live food. Phasmid leg spines are not raptorial and the eyes are much smaller.

Cockroach with ootheca, *Eurycotis opaca*.

Mantisfly.

Mantisflies (Order Neuroptera: Family Mantispidae) look very similar to mantids, but only as adults. Mantisflies are different from mantids in that they have clear membranous wings and go through a complete life cycle which includes a very different looking larva and a pupal stage. Mantis shrimp are neither shrimp nor mantis but are named for the insect they resemble. Mantis shrimp, Order Stomatopoda, are fantastic marine crustaceans from the Class Malacostraca, which also includes true shrimp, lobsters, crabs, and isopods. They are solitary predators armed with raptorial forelegs and huge, compound eyes which project off a head that can be turned all the way around. Even those partial to the terrestrial mantis must admit certain mantis shrimp species are the most extravagantly colored organisms on Earth.

In 1758 Carl Linnaeus placed the first described mantis species in the genus *Gryllus,* at the time a generic name used by Linnaeus for various orthopteroid insects, and in the subgenus *Mantis* along with the phasmids. For more than two centuries, mantids remained part of the Order Orthoptera along with the crickets and grasshoppers, walkingsticks (Order Phasmida), and the mantids' closest relatives, the cockroaches (Order Blattodea). If

A phasmid, *Megaphasma denticrus* male.

one considers the extreme morphological and developmental differences between families within other orders like Coleoptera and Hemiptera, the elevation of the various Orthopterans into orders seems highly irregular. For a time the Mantodea and sister taxon Blattodea were considered suborders of the Order Dictyoptera, Superorder Orthopteroidea (Arnett 1993), but other schemes placed the two orders in their own superorder, the Blattoidea (Stanek 1969). Today the pair are most commonly listed as orders in the Superorder Dictyoptera.

Mantids are separated from cockroaches because they have raptorial front legs with femoral brushes, the head has three rather than two ocelli, and oothecae are cemented to objects during formation. (Cockroach oothecae are formed at the end of the abdomen and later cemented, retracted, or buried.) If an orangehead cockroach *Eublaberus posticus* is observed from the side as it devours a caterpillar, the close relation is apparent. Until recently there was debate as to whether termites were closely related to cockroaches, but genetic tests have resulted in the conclusion that termites are simply social members of the Blattodea (Inward et al. 2007).

Class Insecta
    Superorder Dictyoptera
        Order Blattodea (*Cockroaches and Termites*)
        Order Mantodea (*Mantids*)

There is much work to be done in mantis taxonomy. In 2002 there were 2,300 recognized species, while only ten years later and despite various revisions there are around 2,450 species (Ehrmann 2002, 2011). The Mantodea is divided into fifteen families, though they are not as distinct as one might imagine. (The families had been cut down to just eight near the end of the last century and some of the more unique types have been moved back to their own families.) The largest, the Mantidae, at that time included a jumble of at least eleven different historical families converted to subfamilies; even now some members would still be in the Hymenopodidae if a dichotomous identification key were used. It can be easier to look up what family a genus belongs to than to look at morphological features. Below is a quick look at the various extant families, beginning with the largest. (Family species tallies partly reconciled with Ehrmann (2002), Larsen (2007), Svenson & Whiting (2009), and Otte et al. (2012).)

FAMILY MANTIDAE, the largest by far with around twelve hundred species, contains such animals as the common African mantis *Sphodromantis lineola*, the North American Carolina mantis *Stagmomantis carolina*, and the giant Asian mantis *Hierodula membranacea*. Nearly all the mantids with a standard appearance come from this family. There are, however, many oddities like the Brunners' mantis *Brunneria borealis, Yersiniops* spp. grasshopper mantids, and the amazing deadleaf mantids of the genus

Typical Mantidae.

Family Hymenopodidae. *Pseudocreobotra wahlbergii*.

Family Hymenopodidae. *Otomantis scutigera*, subadult female`s face.

Family Thespidae. *Thesprotia graminis*, large nymph.

Deroplatys. As a group they are known for triangular heads and relatively large, rounded eyes. Nevertheless, a number of genera have rounded heads or have conical eyes often associated with the next family.

Family Hymenopodidae is the second largest with two hundred and twenty-three species and is a widely variable group. It includes species that resemble flowers like the beautiful, tropical Asian orchid mantis *Hymenopus coronatus*, the spiky and extravagant African spiny flower mantids *Pseudocreobotra ocellata* and *P. wahlbergii*, but also the various boxer mantids, and the ghost mantis *Phyllocrania paradoxa*. Members of this family are often colorful and possess lobes on the middle and hind legs. Many have conical eyes. They are differentiated from the Mantidae primarily by the short, laid-down spines on the outer side of the protibia.

Family Tarachodidae has two fewer species than the Hymenopodidae and is represented in the U.S. only by the introduced *Iris oratoria*, while representatives of a few African genera have shown up in captivity: *Pseudogalepsus*, *Tarachodes*, and *Tarachodula*. Females of a number of genera are known for ootheca guarding behavior and possibly maternal care of nymphs.

Family Thespidae includes numerous genera containing a hundred and ninety-two species from across the globe, with most species occurring in South America. Our *Oligonicella* and *Thesprotia* are part of this family. Adult female *Thesprotia graminis* look much like small *Brunneria*, but that genus is in the Mantidae.

Family Iridopterygidae includes a hundred and twenty-five species that are rarely kept, including species of *Tropidomantis* and *Sinomantis*. They are small and the internal organs can be

Family Mantidae. *Phyllovates chlorophaea* nymph.

Family Hymenopodidae. *Creobroter elongatus*.

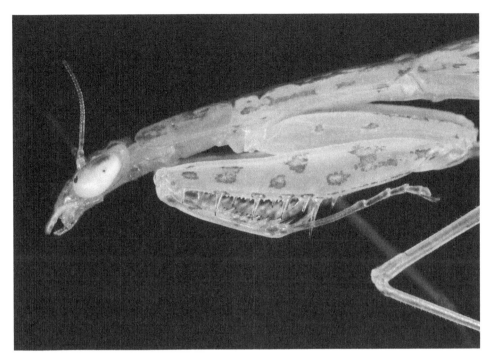

Family Iridopterygidae. *Sinomantis denticulata*. © Tammy Wolfe

Family Sibyllidae. *Sibylla pretiosa* subadult female.

visible through the exoskeleton, leading to common names like glass mantis.

FAMILY ACANTHOPIDAE includes maybe a hundred species found across South America. Members of the genera *Acanthops, Decimiana,* and *Stenophylla* are spectacular leaf mimics, while *Acontista* species look like a flower mantis. Whether they resemble dead leaves or flowers, representatives hold their heads forward of the raptorial forelegs, resembling the demeanor of a mantisfly.

FAMILY AMORPHOSCELIDAE has eighty-four Old World and Australian species but none of these have been cultured for long. They are generally under an inch and have very reduced foreleg spination.

FAMILY EREMIAPHILIDAE has only two genera but around seventy species. The various species look similar, are stout and long-legged, and camouflaged well on rocky terrain. In contrast to other mantids they run very fast. Their habitat is the deserts of Africa and the Middle East.

FAMILY LITURGUSIDAE, with sixty-five species, includes some strangely adapted mantids that live on tree trunks, including our North American *Gonatista grisea*, South American *Liturgusa*, and Asian *Theopompa* species. Representatives are found throughout the world's tropics.

FAMILY TOXODERIDAE has fifty species including some spectacularly strange-looking creatures

Family Tarachodidae.
Pseudogalepsus nigricoxa female.

Family Iridopterygidae. *Tropidomantis* sp.
ex. Thailand. © Kenneth Tinnesen

Family Liturgusidae. Juvenile *Theopompa* sp.
ex. Taiwan. © Yen Saw

Family Acanthopidae.
Leaf-mimicking mantis (*Acanthops* sp.) from Ecuador.
(CC) Geoff Gallice

covered in leafy projections like a leafy dragon seahorse. Members of this Old World family are extremely long and thin with flattened cerci.

FAMILY EMPUSIDAE includes the peculiar wandering violin *Gongylus gongylodes* (Asia), the mythical devil's flower *Idolomantis diabolica* (Africa), and the horned *Empusa pennata* (Europe). The key feature is that males in this family have feathery antennae. They usually have lobes on the legs and cannot climb glass. There are three shorter spines between the large inner spines on the profemur, instead of one. The head is usually diamond-shaped and topped with a large spiky projection. There are barely thirty species and a few dozen subspecies.

Family Empusidae. *Idolomantis* threat display.
© Kenneth Tinnesen

FAMILY SIBYLLIDAE is a small African family with just three genera and sixteen species. The cryptic pretty mantis *Sibylla pretiosa* is a commonly kept species from this family.

FAMILY MANTOIDIDAE is a tiny family of eleven miniature mantids including our native, ant-mimicking *Mantoida maya* from Florida.

FAMILY CHAETEESSIDAE has six tiny, South American species with extremely long cerci. (The mottled nymphs almost look like firebrats discussed in the prey chapter.)

Family Eremiaphilidae. *Eremiaphila* sp. nymph.
© Kenneth Tinnesen

Family Mantoididae. *Mantodea maya*.
© Tony DiTerlizzi

Family Metallyticidae. *Metallyticus violaceus* nymph. © Myke Frigerio

FAMILY METALLYTICIDAE includes just five species in one genus from southeast Asia. These average a little bigger than *Mantoida* and have a short squat pronotum and elongate first outer spines on the profemur. Their claim to fame is their unreal metallic coloration, though they live on tree trunks and behave similar to the Liturgusidae.

Now that you think you have a general grasp on the present understanding of basic classification of mantis species, this chapter ends with a warning. A large project has begun at the Cleveland Museum of Natural History in Ohio to study molecular phylogeny of world mantids. The collection of specimens being amassed for this project will be among the top three in the world. Initial genetic tests (Svenson & Whiting 2009) have led to the discovery that similar looking species may not be necessarily related, and it appears many species grouped together by traditional systematists based on morphological characters have convergently evolved into similar ecomorphs. A variety of species considered bark mantids have proven to come from at least seven different groups (Mangels 2012).

MANTIS BIOLOGY

Prey capture has been the main topic of study for this group of insects. Though the literature seems extensive it focuses on limited experiments that measure the interest in variables such as prey items of varying size, shape, color, background color, speed, and direction of approach. Only a few species have been thoroughly studied and only under laboratory conditions. Even if capture response details are available for the species being kept, it plays almost no role in captive husbandry parameters.

Excellent vision is a hallmark of the mantis. The huge, compound eyes are feats of engineering that detect shape, color, and movement. Since vision is integral to prey capture in this group there are no cave species. If the eyes are painted over, they starve. Feeding occurs at night and in the dark, but only if food runs into the raptorial forelegs. In many species the eyes turn dark at night and return to normal coloration during the day. This transformation takes a few minutes at the least and allows the eyes to absorb more light during periods of low light. The compound eye (made up of ommatidia) has a black spot in the middle which appears to move as the viewer moves. This effect, commonly called the pseudopupil, results from light absorbed by facets of the eye at varying angles. Since every ommatidium is shielded by pigment cells, most of the light that does not reach the ommatidium along its axis is reflected. The ommatidia whose axes are in line with the viewer's do not reflect light towards the viewer, causing those ommatidia to appear black. There is a dedicated patch of comb-like setae on the inside of each profemur, known as the femoral brush, used to keep the large compound eyes clean. Mantids have five eyes, as there are three large simple eyes (ocelli) between and above the compound eyes. These accessory eyes are often larger and more visible on males.

Raptorial forelegs are as important for hunting as superb eyesight. The profemur and protibia are outfitted with spines and fold up like a trap. As mentioned, the placement of these spines has played a major role in classification schemes. Although the front legs are superbly adapted they have the same general structure as the walking legs: coxa, trochanter, femur, tibia, and tarsus. The four rear legs are generally known as walking legs because they are not used in hunting, though without them the front legs have no base and cannot function.

Most species have euplantulae on all six tarsi. These adhesive, deformable pads on the feet aid in climbing smooth surfaces. These structures are why mantids have no trouble walking up smooth glass or plastic surfaces in seeming defiance of the law of gravity. Trace amounts of a complex fluid emulsion and

Night eyes.

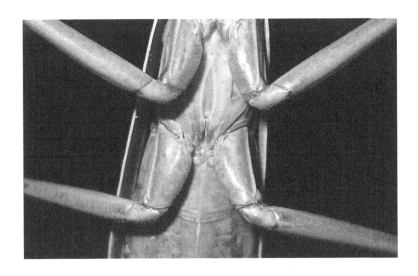

Mantis ear (seen as a deep groove between the base of the hind legs).

Phyllovates chlorophaea face, showing the three ocelli around the horn between the pair of large, bulging eyes.

deformation of the fibrous pad to fit the contours of a surface (Casterena & Codd 2010) allow them to scale nearly any object. An extremely thin layer of oil or petroleum jelly usually prevents traction. Euplantulae use an active rather than passive means of adhesion, so they fail in older mantids and during molting periods. Euplantulae may provide enough adhesion for tiny, light-weight nymphs to molt from almost anything. However, larger animals must hang from their hook-shaped claws when the animal and blood pressure is occupied with the important task of molting. Empusids, ground, and bark mantis cannot climb smooth surfaces because the euplantulae are reduced in size.

Mantids smell using the antennae and taste with the palps. Olfactory sensilla of various types are located on the antennae, which allow the mantis to smell. Odor detection provided by the antennae may allow a mantis to locate better hunting grounds near rotting fruit, certain plants, or a carcass surrounded by flies. Prey is probably never directly located through smell, but females are. Males are commonly outfitted with much longer and more intricate antennae that allow them to pick up the scent of females from long distances. The bipectinate, feathery antennae of male empusids and extremely fuzzy antennae of some male hymenopodids like *Phyllocrania* allow the male to pick up on a female's scent from just a few distant molecules. Two pair of sensory palps on the mouth taste the prey, may aid in odor detection, and are used to guide and direct food while chewing.

Hearing is probably one of the more peculiar aspects of mantis biology. A single ear is located under the body between the base of the rear legs (Yager & Hoy 1986, 1987; Yager & Svenson 2008). Inside a deep and narrow groove is a pair of tympanic membranes. Many neotropical species and flightless females do not have an ear or have only a reduced one. Research has shown flying mantids that possess ears will drop to evade bats when they detect echolocation calls (Yager & May 1990; Yager et al. 1990; Yager 1999; Triblehorn & Yager 2001, 2005). However, mantids had ears 120 million years ago but bats have only been around half that long (Mangels 2012). Species in the genus *Creobroter* have a double ear, one detecting high frequency and the other detecting low frequency sounds (Yager 1996).

The circulatory system is open and has no relation to breathing, which together explains why animals missing parts of the body or the head can sometimes continue on for days or weeks. Massive blood loss and loss of pressure do not rapidly shut down every other body system like they do in vertebrates. Breathing through a trachea system that opens in spiracles is passive and requires no single organ to run. Trachea branch minutely until each cell is connected to the atmosphere. One disadvantage of the trachea system is it is primarily adapted for either excessive or limited airflow. There is some adaptability but it is not instantaneous and falls within a narrow range, so mantids always fall over dead if there is restricted airflow combined with humidity for too long. (The opposite is true for millipedes and many beetles that require little ventilation and desiccate rapidly.) The tracheal system is not always passive in mantids. Large, overweight, females are commonly observed rhythmically contracting the abdominal muscles to improve the flow of air.

The mantis digestive and excretory system is pretty standard for insects. Unlike most invertebrate predators which leave behind the exoskeleton of prey items, mantids chew it up and later eject the undigested chitin in dry frass pellets mixed with crystallized uric acid from their malpighian tubules. Mantids do not

KEEPING THE PRAYING MANTIS

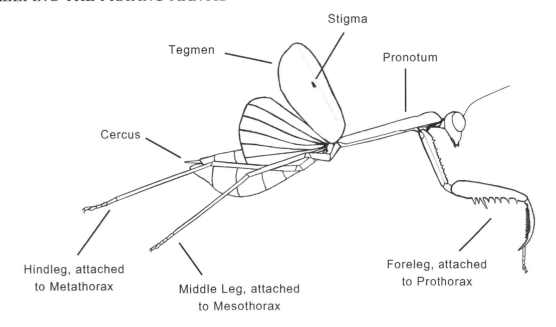

BASIC ANATOMY

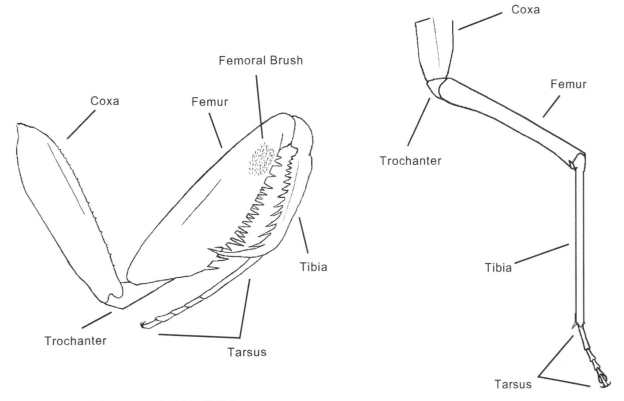

RAPTORIAL FORELEG

WALKING LEG

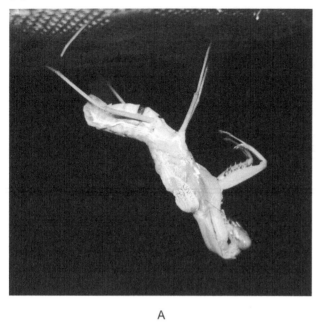

A

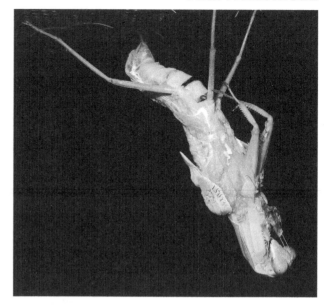

B

C

D

Molt sequence. (A-D)

KEEPING THE PRAYING MANTIS

E

F

G

H

Molt sequence. (E-H)

require flush water for uric acid removal like vertebrates, but they are among the few arthropods to regularly drink as they need to replenish water lost during respiration. Prey fluids and fat content reduce water requirements, but no species are known to have a mechanism for production of metabolic water.

Tegmina are the toughened forewings that hide and protect the larger, membranous hindwings folded beneath. The forewings usually shield the delicate abdomen and are well-camouflaged to help the mantis blend in. Rarely, as in the tiny tegmina of *Brunneria borealis* and stone mimics, do they contrast the body color. Flight capability is often restricted to the male, even in species where both sexes are winged, since females soon after adulthood consume too much food to fly and become heavy with eggs. Many males are superb fliers that often sacrifice camouflage for large wings.

The exoskeleton is made of chitin, which is a long chain sugar polymer similar to cellulose. Chitin can be hardened with the protein sclerotin to create thick, protective body armor. The elastic pleural region of the abdomen can stretch like a balloon but the hardened head, tergites, sternites, and appendage segments cannot stretch. The exoskeleton must be replaced occasionally as the animal inside grows.

The molting process is preceded by the premolt phase and followed by sclerotization or hardening so the entire process can take days, during which time the mantis will not feed. The time taken is not always consistent even for the same species and instar. During premolt the old exoskeleton is partially reabsorbed and can become gray or discolored. The molt itself takes five to twenty minutes as the animal carefully extracts its body from the old exoskeleton and then as long as an hour to inflate the new exoskeleton. Body fluids expand the segments and wings while air is draw in and subsequently released to shape the abdomen and body. Next, in a process known as sclerotization, the larger, soft, pale exoskeleton slowly hardens and develops color. Most of the rigidity develops within thirty minutes of the exuvium being cast off, though a day or more may be needed for large specimens to achieve maximum rigidity of the exoskleton. Sclerotization is a chemical process unrelated to ambient moisture or there would be no aquatic insects. (Misting with water will not slow or reverse the process.) From beginning to end, the entire process (time spent not feeding) is usually one to four days, but the final molt can take much longer. Usually the smaller and younger the animal, the shorter the time period. Small nymphs can often feed the same day.

Praying mantids do not eat the shed exoskeleton like a number of related insects. Many omnivorous cockroaches, vegetarian walkingsticks, and predatory katydids eat the exuvium immediately following a molt but quite a few do not. The exoskeleton has little usable nutrition and since mantids are always exposed, the risk of feeding on the exuvium may outweigh any benefit. (Predators that molt in protected chambers like centipedes immediately consume their old exoskeleton, though tarantulas do not.) The predatory amblypygids have exposed molts and do not feed on the exuvium while Opiliones do. Predatory freshwater crabs (*Geosesarma* sp.) with exposed molts eat their old exoskeleton but must wait a few days after the molt. Feeding or not feeding on the exuvium cannot be expressly associated.

Mantids commonly molt six to nine times between hatching and adulthood. The number is nearly always consistent for individuals of the same species and gender. Some arachnids and most crustaceans continue to molt after reaching sexual maturity, but only the rarest of insects molt after reaching adulthood and none of them are mantids. Any mantis with wings is an adult and will never molt again, though not all mantids have wings when they reach adulthood.

Mantids have only three primary developmental stages: egg, nymph, and adult. Many familiar insects have four life stages: egg, larva, pupa, and adult. Insect larvae look very different from adults and often eat different foods. Flies, mantisflies, butterflies, ants, beetles, etc., change radically in appearance transforming from larva to adult. Mantids, cockroaches, grasshoppers, and true bugs have immature stages called nymphs that resemble a small adult without wings. Because larval beetles, flies, and butterflies look very different from adults, they are named differently: grubs, maggots, caterpillars, etc. The immature stages of mantids, grasshoppers, and true bugs look similar to adults and do not have different names. A nymphal mantis is simply a mantis.

Leg regeneration and autotomy may seem more appropriate to a book on phasmids, but mantids have even more impressive regenerative powers. Of course regeneration only follows molts and will not occur on mature animals. If a mantis nymph that is not already very close to a molt loses a walking leg, the new limb will nearly always be fully functional after a single molt (see photo *Phyllocrania,* also pers. obs., *Hierodula, Phyllovates, Sphodromantis,* and *Stagmomantis*). The leg will be smaller but functional, whereas in other arthropods including most phasmids, two or more molts are normally required for a leg to grow large enough to be ambulatory. Rarely legs do not grow back, but that is related to the point of attachment where the leg was lost. Damaged legs are often autotomized to the point between the coxa and trochanter for regeneration and usually separate there when ripped off. (Autotomy is controlled by nerve bundles and does not occur on

Yellow orchid mantis inflating wings.

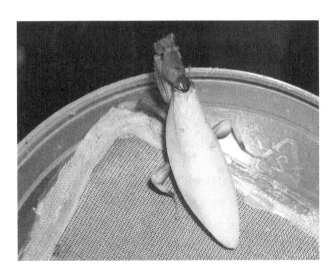
Wings perfectly inflated.

anesthetized animals.) The raptorial forelegs can also grow back but do not regenerate very well and their regeneration can take numerous molts and cause molting problems. Difficulty results because the raptorial forelegs do not always autotomize to a predetermined point when damaged and grow back from the point of damage. If more than a single leg of either type is lost the animal has limited chances of surviving or molting correctly.

Many species are difficult to see in natural habitats. This resemblance to background objects is commonly referred to as *crypsis*. The

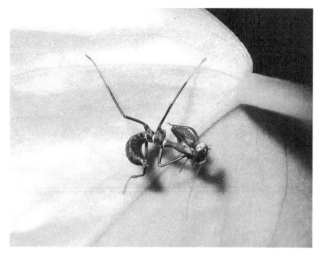

Phyllocrania paradoxa hatchling resembles an ant. Various unrelated ant mimics have white bands on the abdomen which likely resemble the ant pedicel at a distance.

unseen mantis may avoid predation by a bird or small mammal (or people in some parts of Asia, South Africa, and South America) and can ambush prey with ease. Representatives of various genera from the tropical Asian *Deroplatys* to the African *Phyllocrania* and South American *Stenophylla* are unbelievably accurate mimics of dead leaves with every veinlet and wrinkle mirrored. Numerous species from genera like *Choeradodis*, *Orthodera,* and *Rhombodera* have expanded pronota of varying extremes and bright, leaf-green coloration to blend into the foliage. Bark or lichen mantids such as New World genera *Gonatista* and Old World *Theopompa* are mottled and speckled to blend into the lichens covering the bark they live on. There are no North American mantids that look like twigs, but the Old World *Heterochaeta* and *Popa* species are impressive stick mimics rivaling many phasmids. Grass mantids like the American *Thesprotia graminis* and *Brunneria borealis* are nearly impossible to spot among tall grasses. Various European and African empusids add a tapered horn that conceals the head shape among grass while the Asian *Schizocephala* holds thickened antennae forward like horns to maintain the grassy shape. Ground-dwelling members of the American *Litaneutria* and African *Eremiaphila* are lumpy in shape and colored so they blend in with the pebble-covered ground. Various *Hierodula, Mantis, Sphodromantis, Stagmomantis*, and countless other genera with a standard body shape can take on various hues of green and brown and may be quite difficult for predators and humans to locate among the vegetation.

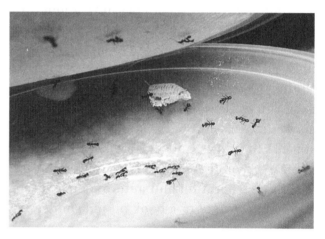

Odontomantis planiceps hatchlings are ant mimics. © Yen Saw

Other species resemble flora or creatures that are common, but not a major component of the general background. A few flower mantids look like flowers, though the majority resemble flower parts, and do not look like an entire flower unto themselves. Hatchling mantids are often easy to see but resemble common insects that larger predators are used to seeing but not eating. Hatchling nymphs of genera as different as *Parasphendale* and *Phyllocrania* look like a swarm of black ants but spread out and take on cryptic colors after the first molt. Some small species like the ant mantis *Odontomantis planiceps* continue the deception for a few instars. Our native *Mantoida maya* look like *Camponotus* ants as they scurry

A grass mimic *Schizocephala bicornis*. © Yen Saw

Dead leaf mimic. Female *D. angustata*.

across the forest floor in nymphal stage, while adults may resemble wasps (Deyrup 1986). According to Tomasinelli (2002, 2003) the long-legged, red-and-black colored hatchlings of *Hymenopus, Parymenopus,* and *Theopropus* probably resemble specific venomous assassin bugs (Hemiptera: Reduviidae) found in southeast Asia. The model these species mimic is uncertain, but the colors, lanky shape, and gait do not suggest ants.

The importance of camouflage in prey capture versus predator avoidance is unknown and is unlikely to be the same for different species and regions. However, many mantids spend hours feeding, grooming, or signaling others through movements of the abdomen or femora during the day. Most cryptic insects are active only at night. The numerous mantis species that are colored to match flowers suggest prey capture could be the more significant consideration. Phasmids whose mimicry is purely defensive never mimic flowers, while a number of predatory assassin bugs resemble flower parts and predators like the crab spiders can change between white, pink, and yellow to match the flowers they hunt on.

Most if not all mantids are able to change their body color in order to adapt their camouflage to the variable environment. Unlike some lizards and cephalopods which can rapidly change coloration at will, mantis color transformations take a long time. Most mantids only change colors immediately following a molt. Different species across various families are

Hymenopus 1st instar nymph. © Kenneth Tinnesen

Orchid mantis subadult female, a flower mimic.

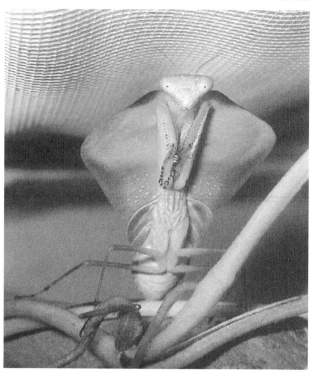
Leaf mimicry in a hooded mantis, *Choeradodis rhomboidea*. © Tom Larsen

The cryptic *Gonatista grisea* female.

Family Empusidae, *Hypsicorypha gracilis*, a grass mimic. © Myke Frigerio

KEEPING THE PRAYING MANTIS

Stick mimicry in *Popa spurca* nymph.
© Tom Boeije

Stick mimic, a young *Heterochaeta* sp.

able to change between green and brown or at least from pale tan to dark brown. Cues taken prior to the molt are considered, so if captive conditions are reversed prior to a molt the result is a halfway or partial color change. Many of the hymenopodids that live on flowers can change from white to purple, pink, green, or yellow without molting, to match flowers or perches in captivity. (It has to be the perch, they will not simply change if a particular color is nearby. Even if there is only one colored perch the animal recognizes color and often sits on the floor of the cage to avoid a perch that is the wrong color (pers. obs.).) Some species found in areas where fires are common develop a percentage of black individuals following brush fires or after burnt sticks are placed in the cage for climbing. These include the African *Phyllocrania paradoxa* and *Pseudogalepsus nigricoxa*. Specimens of *Litaneutria minor* from the desert southwest are often solid black when found near old campfires sites. Even minor color changes take two days to two weeks of exposure, so habitat changes midstream can result in partial or no changes. Hobbyists are often frustrated when they throw a black stick or red piece of paper in the cage and the mantis does not change, but the process is almost never so simple (though it can be for specific species and cues).

A complex set of stimuli and responses allow praying mantids to change colors. Transformations are affected by the coloration and texture of nearby objects (i.e., plants, caging, flowers, screen), nutrition, humidity level, temperature, light intensity, and other environmental variables. Color changes may also be affected by genetics, as one or two individuals are often a different color from their siblings despite being kept under identical conditions. (However this is also true for parthenogenetic phasmids, so the genetic switch can go either way if cues do not lean strongly in one direction.) Green coloration in *Phyllocrania paradoxa* is a sex-linked genetic trait and environmental, since green male nymphs revert at the ultimate molt while females can stay green under specific conditions. The exact stimulus is often unknown because the change can be in response to more than one variable. Humidity and perch coloration are the most important but then there are often multiple perches and major shifts in humidity under captive conditions. When *Parasphendale agrionina* was

These sibling, female *Parymenopus* nymphs were both greenish white, kept on silver aluminum perches (individual 16oz. cages). Perches were replaced with green and purple plastic as shown. The specimen provided the purple perch began to show pink highlights after forty-eight hours but took ten days to become this pink (no molt). The specimen provided a green perch did not change.

After a molt the pink specimen has grown even darker and the specimen on the green perch has stayed white. Following this, the white specimen was given a purple perch. Within 2 days some pink highlights were observed and after another ten days it was as pink as the specimen depicted in the first image.

Final *P. davisoni* adult coloration. This female kept on the purple perch for two weeks following the ultimate molt retained only a hint of rose on the leg petals while the female moved to a green perch after the molt retained no pink. Standard adult coloration is often more stable than that of immatures. The glowing yellow color of adults of this species cannot be done justice in photographs.

imported regularly in the 1990s many of the adults were neon green but in captivity their offspring were always gray. *Brunneria borealis* follows a similar pattern except that for at least fifteen years all evidence I observed suggested the pink captive-reared adults with purple wings could almost always be differentiated from the wild green specimens with pink wings. Recently I met someone whose captive conditions sometimes result in natural looking specimens. (Conversely, R. Trone reports seeing some wild specimens in the pink range.) Some *Rhombodera* nymphs will change from green to brown or yellow when kept dry, but on the molt to adulthood they usually, if not every single time, return to green under the same conditions.

For unknown reasons a handful of species change from brown to green a few days to weeks after the molt to maturity. *Sibylla* adults molt out with brown wings and after about two weeks the wings change to green (McMonigle 2012b). This color transformation is not affected by external stimuli as long as the mantis stays alive. Various body parts of *Blepharopsis mendica* (pers. obs.) and *Idolomantis diabolica* (Schwarz et al. 2009) are brown after the final molt but change to green over the next week. In *Idolomantis* the intensity of the transformation and final coloration is related to humidity, but the transformation itself is a constant. However, in *Blepharopsis* a female may wait as long as four weeks to change color in response to temperature or lighting conditions (T. Wolfe, pers. comm.).

Even the smallest mantis is gargantuan compared to the smallest moth, beetle, cricket, springtail, isopod, etc. The smallest are some male *Mantoida* that can be less than 1/2". The largest is debatable but is one of a few Old World stick-like genera that can exceed six inches. Small animals are the most sensitive to

Pseudocreobotra nymph.

Pseudocreobotra.
Large nymphs can be white or pink.

Idolomantis diabolica, fresh brown male.
© Kenneth Tinnesen

Idolomantis diabolica change to green after two weeks. © Kenneth Tinnesen

shortages of food, water, and moisture because of the tiny body mass, although humidity requirements vary more widely by species than size. Large animals require more prey and larger habitats, and are more likely to compete with or become prey to vertebrate predators.

It is difficult to say how captivity affects adult size. A few species do not grow nearly as large as they do outdoors. This is particularly true for captive *Tenodera* species that often end up half to three-quarters the weight of wild specimens. Nutrition is one variable as specimens of other species fed only with crickets can be stunted, while offering prey dusted with vitamin/calcium powder under the same conditions can improve size. Climatic factors might play a part since adult size seems to be affected in temperate zone species kept indoors at temperatures higher than outside (particularly at night).

Another factor is the cage size. Extremely large cages seldom have an effect but tiny cages can stunt the adult size. (Note, however, a mantis does not "grow" to the size of the cage, so a large species will mismolt and die in an excessively small cage.) Other species can grow as large or larger than wild-caught. My successive generations of *Otomantis scutigera* were nearly identical in size to the original wild female. *Parasphendale agrionina* would average the same size as wild-caught, but a small percentage of captive-reared specimens would grow much large than anything I saw among the imports.

Development tends to be rather rapid as most species grow from the first instar to adulthood in a few months. Even the slowest growing species rarely take over a year to grow to maturity even if purposefully underfed to accomplish this feat. In temperate areas they only have five to six months to develop, mate, and produce a full compliment of oothecae.

KEEPING THE PRAYING MANTIS

Adult longevity can be as much related to the individual as to the species and is influenced by care and habitat. In captivity, males generally live six weeks after their final molt and females twelve weeks. If nymphs were reared quickly or slowly the adult life is generally unchanged unless the rearing conditions result in deformed or unhealthy adults. Females of some large species like *Sphodromantis lineola* can survive nine months, while the only species I am certain can live a full twelve months is *Deroplatys desiccata*, but after five months the females no longer feed heavily or display any reproductive behavior.

As mantids reach the end of their lifespan the males of most species, especially males of small species like *Thesprotia graminis* and *Creobroter gemmatus*, tend to drop dead without warning. The females usually show signs of aging long before they die. Females of larger species, like *Tenodera sinensis* and *Hierodula membranacea* display signs of aging weeks or months before they finally expire. The most common signs of aging include increasing discoloration of the exoskeleton and the wing tips and tarsi growing brittle and breaking off. Old specimens stop climbing high in the cage and may lose the ability to climb or even maintain an upright position.

Three very large, obvious ocelli between the antennae on *Sibylla pretiosa*. © Kenneth Tinnesen

MANTIS BEHAVIOR

Behavior is not easy to separate from biology since physical responses can be difficult to discern from more complex, "software"-driven reactions. Different temperatures can affect some behaviors. In order to isolate a factor responsible for a certain aspect of biology or behavior, it is important to keep other variables constant. Most studies of mantis responses have been limited to very specific stimuli and in most cases the conclusions drawn cannot be extrapolated out to explain any other behaviors because the testing methods considered few variables and forced their subjects into one of two results. In the past, people who studied mantids seemed to think of them as simple robots, but continued discoveries show mantis behaviors are more complex than previously imagined (Prete et al. 1999). Mantids exhibit a number of interesting behaviors of varying complexity that fascinate the enthusiast or offer insight into husbandry parameters.

Individuals behave differently under the same conditions. Each species, gender, and developmental stage may have a different usual response but there will be the odd creature that is terribly aggressive or meek. Meekness is sometimes explained by a health problem or physical defect, while aggression is a character trait. Wimpy mantids may have been undernourished for too long and are simply dying. It is not rare to discover that an individual that has barely showed interest in food has a bent raptorial foreleg or mouthpart. Still, physical defects do not necessarily result in meekness. I once had a *Hierodula patellifera* eaten from the end of his thorax to the tip of the abdomen by a tank mate. This little head and two arms fed aggressively and also attacked my fingers, which made hand feeding more difficult than usual (of course, he did not survive to the next molt).

Praying mantids watch the keeper and show some ability to recognize individuals though this awareness could easily be a simple response to dissimilar stimuli. They seem aware of humans to an extent that other invertebrate pets either are not, or do not appear to be. They often watch the keeper and follow his/her movement with the angle of their head (not to be confused with the illusion of observation the pseudopupils impart). Often we seem to be rapidly assessed and then shrugged off as the animal continues on whatever goal it has in mind. Mantids do become accustomed to handling and the most aggressive specimen eventually grows bored with attacking the hand that feeds it (until a few hours later, anyway).

A universal and yet unusual behavior of mantids is that most spend nearly all their time hanging upside-down from the lid. The mantis may walk around the cage or come down when food is offered, but later ends up back under

The pseudo pupil is not visible from this direction. © Kenneth Tinnesen

The visible pseudo pupil imparts a sense of personality and intelligence. It is an illusion—this is the same animal as above from a different angle. © Kenneth Tinnesen

Mantis showing pseudo pupils. © Kenneth Tinnesen

the lid. Some run around constantly and seem very unhappy if they are not allowed to get to their upside-down perch. When a mantis descends to catch an insect on the floor it may back up and hang upside-down to eat (though I have seen some of the terribly voracious species lie on their backs feeding). For large species the heavy body can make hanging less energy-consuming than resting upright. Many species that hang upside-down in nature do so at an angle rather than horizontally. I have seen a number of mantids outdoors and many seem to be sitting in an upright position, but as soon as the same animal is placed in a cage it goes straight for the lid and hangs. Conversely, if a *Rhombodera* is moved from a small cage to a large one with plants, it usually spends its time upright on a green leaf. Ground and bark mantids that normally could not hang upside-down, and cannot climb glass, will climb a stick in the cage and hang upside-down from the lid. These particular mantids cannot hang this way in nature but tend to behave this way if given the chance to climb. This habit is probably the result of a drive to find high ground due to a restricted cage size rather than a desire to hang upside-down.

Hanging upside-down is normal in captivity but in a few rare cases with difficult species or individuals it results in death or deformity. If hatchling *Gonatista* find themselves on a horizontal plane, they often hang uselessly by the rear two legs and starve or dry out. Some *Tenodera* have weak abdomens that flop over and crease. This results in death if the cage is not immediately redesigned. A problem for a small percentage of individuals of various species is, as they undergo the final molt, they try to expand the wings while holding onto the lid with the rear legs instead of hanging down from the front and middle legs. The wings bend downwards as they expand and end up

Polyspilota aeruginosa grooming.

deformed. Why these species or individuals cannot control an obviously harmful behavior is unknown, but similar surfaces are probably never encountered in nature.

Grooming takes up a large part of mantis' activities. The huge eyes are regularly wiped with the femoral brushes to maintain good vision. Praying mantids often lick the femoral spines and front tarsi, pull down the antennae, or shift a rear leg forward so the tarsi can be methodically cleaned of debris. Half an hour can be spent as they carefully clean up the raptorial forelegs following a meal. Beginning hobbyists sometimes notice a missing body part on a mantis while it is grooming and think it is eating itself. The body part was already missing.

Many species react visibly to thirst and hunger. When the cage is misted their reaction is related to how thirsty they are. A thirsty mantis rapidly locates water droplets and drinks greedily, while a hydrated individual will not even bother to look at the water. In captivity, an extreme reaction or lack of response can let the keeper know whether misting should be scheduled more or less often.

KEEPING THE PRAYING MANTIS

Parymenopus nymph grooming
(before the purple perches).

Hunger responses are often less telling. Extremely hungry empusids sometimes do not even move when food is added, while some *Rhombodera* and *Hierodula* will jump at prey and lay on their back greedily feeding despite a somewhat rotund abdomen.

High temperatures can cause mantids to seem energetic and nervous. Adult males readily take flight above 72° F, but at lower temperatures they may need to be sufficiently annoyed or given time to warm up the wing muscles first.

A number of species employ colorful signals to communicate with other members of the same species. These behaviors can seem mysterious if animals are kept individually. The enlarged femora of the various boxer mantis species are usually orange or red with black marks on the inside surface. One forearm is moved forward to expose the bright colors and commonly used to convey identity. (When the femora are held together the outer, cryptic surface of each covers the other.) They probably should be called flagging or flag semaphore mantids, but other species use other parts of their bodies that do not loosely resemble large boxing gloves on a skinny arm. *Creobroter* and *Pseudocreobotra* nymphs flash their brightly marked abdomens at each other in the same fashion. The colorful rear wings of various large species are often part of the courtship display used to identify prospective mates.

Females may arch the abdomen when broadcasting pheromones (Prete et al. 1999; Gemeno et al. 2005; Perez 2005). The wings are held up or straight and the abdomen is curved downward. In *Acanthops* the female inflates sections of the intersegmental membrane on the dorsal surface of the abdomen when "calling" (see photo; interestingly these are the same inflated knobs on *Heterochaeta* and some other nymphs that appear to be used in camouflage). In *Rhombodera* and many others, when the abdomen is arched, segments at the

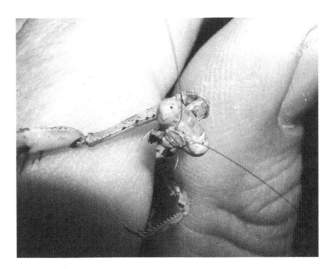

Stagmomantis carolina female
biting the hand that feeds her.

tip of the abdomen used in ootheca construction are intermittently exercised (pers. obs.). This "calling" behavior is very common for older females and usually is only seen on specimens that have not mated. If a mated specimen displays this behavior there is a good chance the female should be mated again with a different

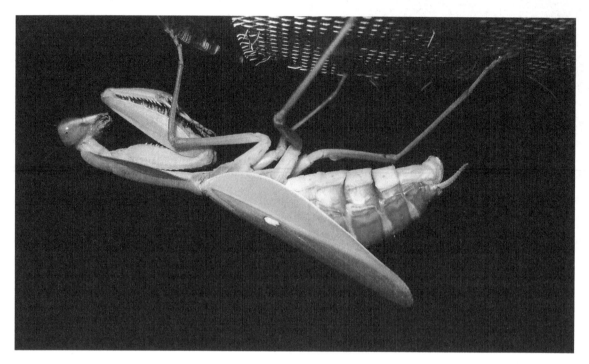
Female calling with pheromones.

Otomantis scutigera displaying brightly marked inner surface of the femur.

male since the spermatophore may have been rejected or not transferred at all. However, after the deposition of an ootheca, the female may start to "call" again, even if she is still fertilized (Berg et al. 2011).

Camouflage must be accompanied by behavior. Dead leaves, chunks of tree bark, and beautiful orchid flowers do not run up and down trees or jet across the forest floor. Species that resemble ants must run around and approximate the jerky movements of ants, or the disguise is of limited value. In the same way, assassin bugs do not run in rapid, stilted, ant-like movements. Dried leaf mimics across various families hold their bodies in strangely contorted shapes so the shape of an insect body is obscured. When they perceive a threat, bark mantids remain still and press their bodies tightly against the bark with legs naturally splayed out to the sides so as not to leave an unnatural shadow. An elaborate disguise will not do much good if it stands off the bark with long rear legs to create an insect-shaped shadow. Elongate species often hold the front legs out ahead of them like a stick insect instead of in the characteristic praying position. *Heterochaeta* often hold the elongate cerci and raptorial legs at nearly right angles from the sides to make the appendages look like small branches, while inflating small, knobby protrusions of the dorsal abdominal intersegmental membrane. Without behavioral adaptations the specialized body is not likely to match the branched, knobby growth of specific plants they are found on. Lastly, they can recognize the color and shape of the background and commonly pick perches in captivity they match best (Edmunds & Brunner 1999). Looking and acting like a rock or lichen will not conceal a creature sitting on a big green leaf. Flower mantids pick matching flowers if given an obvious choice.

Nearly all praying mantis species will display a sort of jerky, swaying movement at certain times when they walk. The primary reason for this peculiar movement is that swaying plant movement is often ignored by both predator and prey (Lasebny & McMonigle 2001; Watanabe & Yano 2009, in press). The mantids seem to sway back and forth like leaves and branches in the wind. When wind speed increases, swaying movements increase accordingly. When prey is not within striking range they may slowly creep up on the intended victim as they sway side to side and front to back. According to other authors this movement also helps the mantis to better judge the distance of objects through motion parallaxis (Poteser & Kral 1995; Kral & Poteser 1997; Kral 1998).

Related vegetarian, camouflaged insects carefully hide all movement and only become active when their movements cannot be viewed. Feeding and dispersal normally take place at night. In mantids, feeding occurs primarily during the daylight hours. Rain covers the movements of cryptic insects and after the habitat is misted, many phasmids and leaf katydids walk around the cage and feed in an apparent response to the cover of "rain." Mantids will drink if thirsty but do not respond to rain as a cue to action.

If camouflage fails, mantids can employ a variety of flight behaviors. Bark mantids like *Gonatista* run sideways with speed and make certain they are never on the same side of the tree as the predator. Many of the ground dwelling species can run very quickly across flat surfaces, as in *Eremiaphila* and *Litaneutria*, or have long rear legs used to jump, as in *Yersiniops*. Some species have colorful areas on the abdomen or hind wings which are flashed as the animal tries to escape and then immediately disappear along with the mantis as the tegmina are closed. Flower mantids like

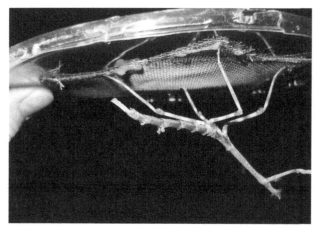

A large bifurcated knob is expanded between segment 4 and 5 and a much smaller knob between 4 and 6. This nymph is two molts from maturity but very small nymphs display the same behavior.

Deroplatys large nymph threat display.

Heterochaeta. When disturbed, expansions of the intersegmental membrane are immediately withdrawn.

Pseudocreobotra tuck the legs in and drop while using the tarsi to grab and push to hit as many leaves as possible on the way down in hopes of being lost in all the movement. Many species try to evade predators by playing dead after a fall, while others like *Acromantis* intermittently play dead and display spastic movements.

The most visibly spectacular of all mantis behaviors is the threat display. When escape behaviors fail most species maintain a defensive position also known as the deimatic display. This threat display is often elaborate as the mantis raises both sets of wings, twists its abdomen, turns to the side to appear larger, and holds the forelegs back and ready to strike. Each species has its own unique defense position. Lifting the wings exposes colorful surfaces under the tegmina, on the hindwings, or abdomen. The mandibles are widely opened and there can be colorful surfaces exposed. Some species are decorated with eyespots on the wings which, when exposed, look like the eyes of a much larger animal. Eyespots on the wings not only draw attention away from vulnerable body parts, but also hopefully startle the aggressor so that the mantis can escape. Brightly colored, normally hidden hind wings and abdominal segments may also serve to startle attackers. A few species include sound in their defense and rub the forewings against the walking legs or the hind wings against the abdomen to make a rasping noise (Crane 1952; Edmunds 1972, 1976; Edmunds & Brunner 1999).

In the past, entomologists often studied mantids without any knowledge of their behavior in real life and made some interesting

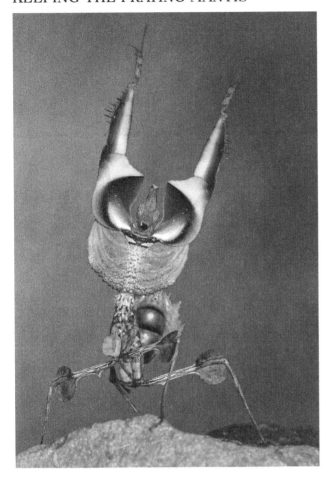

Idolomantis diabolica "flower." © Tammy Wolfe

conclusions. Some of the most famous "flower" mantids of the past were not among the countless species that we now consider flower mimics. The devil's flower mantis *Idolomantis* is the most published example of a threat display confused with a flower (reviewed in Schwarz et al. 2009). The name devil's flower comes from multiple texts around 1900 (and repeated for a hundred years) that suggested it sat around in threat display pretending to be a flower to catch food (Sharp 1899). None of the writers had seen a live specimen and those that wrote about the actual behavior did not have as interesting a story. In Helfer (1963), *Deroplatys lobata* was suddenly turned into an "orchid mantis." Diagrams show the backside of this deadleaf mantis in full threat display compared to an orchid flower drawing. It is a pretty convincing drawing if you do not realize mantids do not sit around in threat display all day and the orchid flower in the drawing is what the entomologist thought the flower it resembles would look like. There is of course no such flower and the mantis looks far less like a flower than the drawing. A final case of a mistaken flower does not appear related to the behavioral threat display. *Gongylus gongylodes* is sometimes referred to as the "Indian rose mantis." The reference dates back to Williams & Sharp (1904) who observed this Empusid in India and noticed the striking resemblance of the purple, enlarged ventral part of the pronotum to a blossom. They state that pollinating insects are attracted to this false "flower" only to fly to their doom.

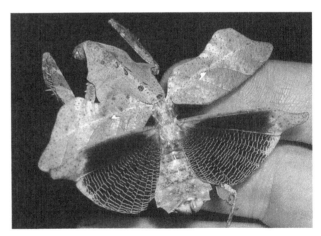

This is the photographic version of the supposed "orchid flower" drawing labeled in the Helfer book.

If the threat display does not deter the predator, the last line of defense is attack. Raptorial legs are held up and ready to deflect small attackers. When a large predator, or the keeper's hand, grabs a frightened mantis, it stabs the attacker with the tibial spines and bites with surprising ferocity. Large species are the most likely to behave this way but even small *Pseudocreobotra* can be rather vicious when attacked. No species have venom and

Threat displays are not always colorful. *Cilnia humeralis*.

Tenodera sinensis threat display.

KEEPING THE PRAYING MANTIS

even the largest cannot do appreciable damage to a human, but the sharp spine on the tibia and powerful mouthparts can inflict surprising pain for how little damage is inflicted. *Cilnia humeralis* and *Rhombodera stalii* often draw a tiny drop of blood when they bite. A fierce physical response may not scare off the predator, but can be adequate to startle or delay the attacker so the mantis can fly off or run down into dense vegetation and escape. Physical attacks lessen or stop altogether when the stimulus is repeated often (such as captive handling).

The majority of mantis behaviors may be "hardwired" into the nervous system but usage is regulated by an ability to learn and adapt to a range of stimuli. There are also behaviors that adversely affect captive care, such as preferences for perches which cause feeding and molting difficulty or behavior that prevents interest in perfectly good prey. The amazing variety of behaviors, from those specifically found in certain species to individual differences, are what make caring for mantids so enjoyable.

Popa spurca threat display. © Myke Frigerio

Threat display. (CC) Marakawalv / Flickr.com

HABITAT CONSIDERATIONS

Popular cage styles include plastic containers, full screen cages, preformed creature cages, and glass terraria (the most common and inexpensive is the 10-gallon aquarium normally used for tropical fish). Plastic containers from the humble deli-cup to the old pretzel jar or massive plastic tote are the most useful, since they can be adapted to any size and style imagined. Screen cages like those for chameleons or the cloth pop-ups can be great for breeding containers and communal cages for large animals or species with minimal humidity requirements. The popular plastic critter cages are functional, but not if they have to be adapted to specific needs, since they are made of a material that is very hard and cracks easily. They are also far more expensive than a deli-cup. Glass terraria offer the clearest viewing surfaces and can last a human lifetime (plastic scratches terribly, grows brittle, and yellows) but by comparison are extremely expensive and the smallest commonly available sizes (2.5 and 5-gallon) are too large for some uses. Vented terrarium lids lined with window screen used to be the only popular vented lid available but large screen and partial hole style lids are used almost exclusively today. The new style lids are useless for containing tiny animals and can be difficult for mantids to hold onto. There are other, less popular habitat options—paper bags adapted with cellophane windows may have once been common in some circles (Dunn 1993). Pop bottles with the top cut off and old pantyhose stretched over the top is one way to recycle. Alternatively in the UK, demi-johns could be bought at cheaper boot sales and secured to a record player disc with a pen torch (Symes 1994).

The size of the habitat will depend greatly on the number and type of mantis, what prey is being used, and what the goal is. The greatest consideration on cage size is whether it will be for solitary or communal housing. It may sound strange to consider keeping predatory cannibals in the same cage, but it is relatively standard in mantis husbandry, especially during the initial instars. Post-separation losses can be much higher even if it were realistic to provide each specimen a cage similar in size to a large, initial grow-out chamber, because there are only twenty-four hours in a day. With large hatches it is simply not realistic to feed and water each specimen separately, due to time constraints and prey escape. The majority of species tolerate each other to a point as long as there is adequate food and space. Small nymphs of the same size can get along very well, though individuals show up every so often that attempt to eat every tank-mate. *Stagmomantis carolina*, *Cilnia humeralis*, *Mantis religiosa*, *Iris oratoria*, and some *Tenodera* species are rather aggressive, irrespective of

Various captive habitats at MantisPlace.com.

cage size, and should be separated as early as possible. Due to cost and space constraints, the larger the individual cage, usually the smaller the number of individuals that can be kept alive. Another major consideration is the available prey, since smaller caging with less obstructions will allow the use of almost any prey type, while large cages and various decorations require the use of highly ambulatory, non-burrowing prey. Lastly, small habitats are more difficult to decorate or use as decorative room accents.

Keep an eye out for common hazards when choosing and constructing the cage. Sharp edges can damage the abdomen of large animals, while small wires sticking out of metal screen can be equally damaging. If the habitat is to house tiny nymphs, it is very important to check the cracks and edges for house spiders. One very small house spider can wreak a lot of carnage in a communal cage containing a number of early instar mantids. Large cages that have been out of use for some time almost always have house spiders inside. Plants are another hiding spot for house spiders, but if the plants are brought in from outdoors they may also harbor crab spiders, wolf spiders, jumping spiders, and assassin bugs. Check for cracks or loose screening that will allow prey or even small nymphs to escape. Some plastics do not adhere well to common glues and screening can come loose without being obvious. Lastly, look for and clean off greasy surfaces that prevent climbing or glue residue that can trap or cripple small animals.

Helpful tools in cage preparation include a utility knife, hot glue gun, screen, silk flowers, foam, peanut butter, and vented deli-cup lids. The utility knife is used to cut out the vent and feeding holes. The hot glue gun is needed to attach the screen to the vents and perches, if desired, to the sides. Any screen can be used to line one or more walls of the container while metal screen is useful to create removable perches or ladders. Silk flowers can be glued to the upper cage sides for more permanent perches (low perches even in tall cages can result in bad molts). Foam is cut to fill feeding holes and can be almost any shape or size as long as it is a little bigger than the hole it is stopping up. To remove residue, peanut butter can be placed on old stickers and allowed to set for a day, then wiped off with a paper towel (oils, rubbing alcohol, and various goo removal products should also work). Physical methods of adhesive removal normally scratch up the surface and tend to look worse than the sticker did. In recent years, deli-cup lids have become commonly available with screen or coffee filter style vents built in. These make great lids for small cages or can be placed in carefully cut holes in large plastic containers. However, the coffee filter paper makes a poor gripping surfaces for some, mostly large, mantids to molt from.

Although hatchling mantids are not small enough to escape through normal window screen, foods such as fruit flies and springtails are only contained with microscreen (fine metal screening) or thin fabric. Cloth can also be used to line the screen lid of a terrarium to

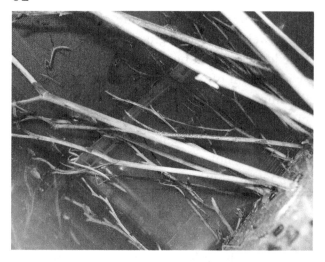

Glued in branches for a large *Tenodera* habitat, at MantisPlace.com.

Tenodera habitat screen lid, at MantisPlace.com.

prevent escape of tiny food. If phorid flies or fruit flies are to be attracted into the cage, window screen should be used, or microscreen can be used on the large vents and a few 1/16" holes drilled on the lower sides. The vents in preformed small animal cages provide adequate airflow but the holes are usually too large and should be covered in screen to offer a better surface for gripping and for containing prey.

Caging can be adapted with screen glued against plastic or glass walls to allow non-flying prey to climb up within reach of the mantids or to let empusids climb easily. Keep in mind, mantids grow weary of trying to catch prey visible on the other side of a screen and can lose interest in the same type of prey even when it is not behind the screen. Cork bark slabs or foam painted or covered with moss or dirt is an alternative for large, decorative display caging. Depending on construction, these adaptations will create large or small places for prey to hide behind. Also, while this aspect of habitat construction improves the mantids access to certain prey, it also improves the access of hungry crickets to a molting mantis.

A feeding hole and plastic funnel are useful items. They will greatly reduce the time it takes to feed, especially when flies are used. Otherwise prey can escape during feeding and small mantids can be smashed when the lid is closed. The feeding hole is usually made near the top on the side and when not in use is plugged with a foam stopper or sponge. The lid no longer needs to be removed, and fruit flies and other insects can be dumped into the hole through the funnel. Remember the funnel has to be tapped to knock down fruit flies, but this is a skill that is easy to master quickly, unless the hole is too low on the cage side. If large flies are being fed in large habitats, the hole is larger and made close to the bottom since the flies fly upward. The feeding hole should be just large enough to fit the mouth of the pop bottle.

Drinking water is usually added at regular intervals. A misting or spray bottle is an important tool. Distilled water (not spring water) is commonly suggested because it prevents build up of calcium deposits on the screen and cage sides. (Calcium deposits can be removed with diluted vinegar in a spray bottle after mantids are removed.) Tap water works just as well and may even offer a slight benefit to the animals since it is not lacking in minerals. (Vitamin water is another option but has little history, while sugary fluids result in a sticky

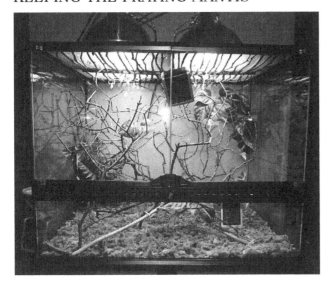

Idolomantis diabolica habitat. 18"x18"x24" Exo Terra with four 23-watt compacts (curly fluorescents) that provide more light and less heat than an equivalent incandescent. The four bulbs raise the habitat temperature 10-12 degrees above room temperature (one compact will do the same for a 12"x12"x12"). An 18-watt heat mat is on the bottom glass for cold nights, or as needed. © Nick Jackson

have more than one vent since cross ventilation is very helpful in providing fresh air. However, if increased ventilation is needed but more vents cannot be added, the airflow can still be improved. A small fan can be added to blow on the cages or fresh air can be directly pumped into the cage with an aquarium air pump. Full screen cages work great, but if the ambient humidity is below 25%, small animals and species sensitive to low moisture will not survive despite constant misting. Also, hard screen and excessive airflow may cause the protarsi of *Gongylus* and *Idolomantis* to dry up and break off prematurely, so cloth screen is recommended.

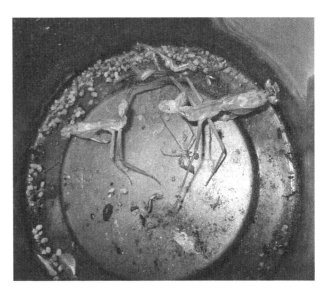

Bare bottom habitats do not require constant cleaning. One individual's developmental lifetime of frass and what remains of prey and molts are visible. Frass production is minimal compared to many animals.

mess.) An eyedropper is a watering tool primarily for use in tiny habitats. Drinking water should be available even if the animals are well-fed, but more so if a feeding is skipped. The inside of the cage should be lightly misted with water once a day, though once a week may be plenty for large animals or certain species and conditions. If droplets do not dry up in thirty minutes, there is not adequate ventilation or too much water was sprayed. High moisture without adequate ventilation will lead to the death of the animals. Misting the cage while the mantis is molting may knock it down and cause a bad molt, but most likely it had gotten stuck in the molt hours earlier and it was not noticed until this time.

Habitats should have adequate ventilation or the animals will fall over dead. Mantis rearing containers usually have at least one 2"x 2" hole covered over with screen. Cages tend to

Ventilation and moisture level work against each other. The more water added, the more ventilation is needed. When I was breeding thousands of mantids for years on end, my method involved limiting both. At 2nd or 3rd instar I would place all the small species and nymphs of larger species in 5oz. deli cups with 2 thumbtack holes in the lid. A ladder (strip of

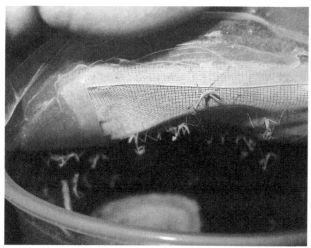

Hatchling *Stagmomantis limbata* in a small growout cage feeding on *Drosophila melanogaster*.

These nymph growout cages have seen dozens of species and countless nymphs.

plastic for yarn work) was cut to fit from the bottom to the opposite side of the top. Depending on the season I would add 1 drop of water to the bottom of the ladder every one to four days and feed every other day. No substrate. Frass must be removed at every feeding and water droplets cannot be too large. If a *strict* regimen is followed the survival rate is near perfect and of course cannibalism is prevented, but the effort required to provide the necessary precision is not likely to be desired by the average enthusiast. (And what does one do with 300-3,000 mantids anyway?)

Species or early instar nymphs that require high humidity can be provided with half an inch of potting soil, sphagnum moss, or other substrate that retains and slowly releases moisture. Sand and gravel do not work well. Coconut fiber holds moisture like soil and has the added benefit of changing from dark brown when wet to orange as it dries. Paper towel is cheap and easy to clean but offers no depth and can look wet when it is dried, which is dangerous. Excessive mold growth on refuse and damp substrate can be harmful, so the use of isopods as a clean-up-crew is often recommended if the humidity is to be retained long-term. It is important to not constantly soak down substrate since too much moisture can result in dead animals. Mantids can adapt somewhat to excessive humidity levels if given time, but they will become fragile and prone to rapid dehydration like plants grown in terraria. Substrate is sometimes a necessity but can make it more difficult to regulate humidity relative to how thoroughly it is watered, than a bare-bottom cage. This can also take practice, like properly watering plants. Growth under artificially high humidity levels can make even larger animals susceptible to rapid desiccation and minor human error (such as forgetting a watering or leaving a lid off), then results in a swift death.

In northern areas the ambient humidity is subject to huge shifts. Home heating and low dew points result in less than 5% humidity in mid-winter and can climb to nearly 100% in spring and fall. The misting or watering schedule should take this into account. It may be necessary to spray the cage twice a day in mid-winter but only twice a week in fall or spring. Summer humidity levels are rarely at the extremes but can fluctuate daily depending on air conditioning usage. A humidity meter will help the culturist understand what is going on since every year is different and allows the misting regimen to be tailored accordingly, but use of small meters inside habitats offers only mixed, useless data.

Heterochaeta cages are much larger than those required for other species.

There are ways to improve humidity in northern areas, but overall a spray bottle (or dropper) in conjunction with proper habitat design is the best choice for cost, simplicity, and usefulness. I found out quickly that a full-home humidifier hooked to the tap to provide endless water for humidification would not work due to something called the dew point. This is the temperature at which water condenses and even well-insulated windows will pull most of the extra water right out of the air and onto the floor below them at an alarming rate when the difference is great. Air from an air pump can be forced through a diffuser in a water-filled container with a tightly sealed inlet and outlet to provide humid air. (Just running air over some water would not greatly affect humidity unless there was quite a bit of surface area.) Such a device could be easily made with cheap containers and gaskets or hot glue but I have never heard of anyone, let alone a successful breeder, who has set up an intricate system that can be prone to equipment failure, require its own maintenance, and may or may not increase the level of attention the mantids will need. Time spent observing and attending to the needs of the animals is time spent more wisely. As already mentioned, substrate and ventilation are important factors for small nymphs of species that require high humidity, but they are limited and controlled by the volume and frequency of moisture addition.

A piece of metal screen or stick should stretch from the bottom corner of one side to the top corner of the other side for the mantis to hang from as it molts. A large piece of screen or cloth glued in as a vent or glued to the underside of a solid lid provides another possible molting surface (paper towel can be used for smaller nymphs). It will not matter how tall the tank is if the molting surface is inadequate or too close to the bottom. As the longer species grow, such as *Tenodera* and *Heterochaeta*, they should be transferred to containers with a molting surface at least an inch or two longer than their final length. Cage decorations must sometimes be removed to force the molt to take place from the screen lid. There should be clear distance from the molting surface to the bottom of the cage that is longer than the length of the adult mantis for it to hang while molting. (This is why a cage height of twice the animal's length is most commonly recommended.) The distance from the molting surface is more important than overall cage height,

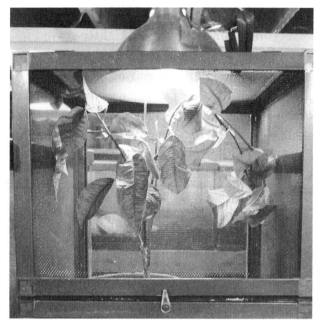

8-gallon screen cage.

while molts are often fine if the clear space is slightly less than the mantis length from the molting surface.

Humidity, molting surfaces, and adequate space are paramount considerations for the habitat for growing mantids. Humidity levels have a minimal effect during the molt even if the air is incredibly dry because the fluids used in molting are already inside the animal. Molting can be dangerous when the air is extremely dry from heating or air conditioning if ventilation is not reduced accordingly, but it is the days leading up to the molt, not during, that matter. Once the molt is underway there is no way to rehydrate the animal's body fluids. Under low humidity conditions it is important to mist mantids a few times a day long before they are ready to molt. As mentioned, the bottom of the container or cage can contain substrate, such as sand or sterilized soil, to absorb excess moisture and keep in some humidity if a problematic molt is expected, but it can retain too much moisture or prevent the animal from adjusting to a forgotten misting. The time when a mantis is about to molt must also be monitored carefully because some live foods, primarily crickets, must be removed from the habitat prior to the molt. An active cricket or house fly can knock the mantis off its perch, causing the mantis to get stuck in its molt or lose legs while molting. Worst of all, a hungry cricket can chew on a mantis while it is in that helpless state. Without enough space, humidity, an undisturbed time period, or usable molting surfaces, the mantis is likely to molt poorly and become deformed or die. If the mantis is already an adult, none of these considerations are important as there will be no more molts.

Caging often changes as the mantis grows:

1. Nymphs from a single ootheca fit nicely in a 32oz. deli cup for small genera like *Phyllocrania, Miomantis,* and *Thesprotia.* One-gallon pretzel jars are useful for larger genera and larger oothecae like *Stagmomantis* and *Rhombodera*. If the first ootheca hatches a small number and a second hatches within three days, they can be placed together. Longer than that, even with less aggressive species, and the second ootheca is just being fed to the first. This initial cage should have 1/2" of soil or coco fiber on the bottom that is kept damp and watered whenever it begins to dry. The screening should be narrow enough to contain small fruit flies. A small chunk of apple placed on the bottom will help keep flies alive. Other fruits should be avoided since they are too sticky or rot too quickly.

2. After the nymphs have molted a few times they can be moved to larger, bare-bottom containers. Movement to individual cages is determined almost entirely

by cannibalism. They are kept together in the large cage as long as they do not harm each other or there are too many to separate. Individuals observed attacking or eating tank mates are removed first since the less violent survivors may be able to remain together a few more molts. They are moved to anything from a 5oz. to 16oz. container depending on size. If placed in individual cages upon hatching, a lot more care is required and more can end up dying because they are more fragile and the amount of time required for hundreds or thousands of nymphs may be impossible to adequately provide.

3. All the small species can stay in 5-16oz. containers till adulthood, but anything more than 2.5" at maturity will need to be moved to at least a 32oz. container a few molts before maturity. They are not moved immediately to larger cages because it is more difficult to feed tiny mantids when the perch does not offer access to the bottom and there is only so much room for large cages.

4. Adult mantids are sometimes housed communally. This requires large, planted enclosures to offer suitable space and hiding places. This method is generally used with less aggressive species and when large numbers of specimens are available.

Habitat temperature affects behaviors and growth but is primarily significant because it affects moisture levels. Higher temperatures (80-90° F) may require the cage to be misted half a dozen times a day. Higher temperatures also allow mantids to grow and molt more quickly. Low temperatures (55-70° F) produce the opposite effect and cause mantids to grow more slowly, stop feeding, or less commonly contribute to a bad molt. Tropical and temperate mantids may survive short periods as cold as 45° F, but appear nearly dead until warmed up.

Lights are often placed directly on cages for empusids. These are often placed over a screen lid on a solid-walled cage. The lights are used to raise the temperature and provide a day-night temperature cycle, but if used over screen cages the heat and ventilation can result in desiccating conditions. Usually only 10- or 15-watt bulbs are used unless the habitat is very large. Do not set the light close to something that can catch fire easily.

Planted cages are acceptable but may require the use of flying prey. Many prey items will dig into the soil and disappear or hide under the leaves. The only really good prey for use in a heavily planted terrarium would be flies or active moths. Crickets may be active enough but often stay near the ground, hide, and only come out at night to chew on the plants or the mantis. Non-burrowing roaches like *Dorylaea* are active and will not chew on mantids, but can find places to hide and may not spend time in the upper regions of the habitat. Various burrowing prey is the least useful since it disappears immediately and may never be caught. Since mantids do not eat plants the source of terrarium flora is not important, though it is not entirely impossible for systemic pesticides to be picked up by prey and then consumed.

There are no hard and fast rules about cleaning the habitat. Bare-bottom cages can be shaken over a trash container before feeding or misting. The volume of frass produced by an adult female is normally more than the volume produced by all the previous stages combined and tripled. Since waste is primarily frass pellets and shed exoskeletons that barely break down, cleaning is done mostly for aesthetics. However, if the cage is low ventilation and low moisture, it must be cleaned at every feeding or the wet, rotting frass will kill the mantis.

Habitat design is important. Lack of ventilation, humidity, or adequate molting surfaces will cause health problems and death. No amount of planning can account for every possibility and the most significant single variable is the care provided. It is important to observe the behavior and health of the animals to adjust feeding and watering and to determine if there are problems with the design.

Stagmomantis limbata free-ranged on a houseplant.

HEALTH ISSUES

The only way to effectively discourage health problems is through careful attention paid to the animals and environmental conditions since few issues can be meaningfully treated or fixed. Most can be prevented almost entirely by a consistent feeding and watering regimen and a proper habitat. However, not all health problems are significant and some are related to old age.

Bad molts are probably the most common cause of death in captivity other than cannibalism. Mismolts can be prevented through proper habitat conditions outlined in the previous chapter and adequate food outlined in a later chapter. If the keeper happens to knock down a molting mantis, it probably was stuck there for a while. If it really is just starting to molt the new exoskeleton will be extremely pale and it will be able to extract itself from the old exoskeleton. In this case it can be carefully hung on a stick or by arranging a few insect pins against a vertical board to support the old exoskeleton. There must be no folded over areas that cause creases and trap appendages. The body should be positioned so it can fall downward and use gravity to help gain freedom from the exuvium. Do not pull or attempt to aid molting because within thirty minutes or so of the molt's beginning, the new exoskeleton is so soft the slightest pressure will tear it or stretch it like taffy (it does not spring back). If a completed mismolt is not too terrible, it may partly or totally heal after the next molt. Once the new exoskeleton is hardened it is okay to carefully pull off pieces of the old exoskeleton to free antennae and legs, but the shape of the molted body cannot be improved. Slightly twisted legs can straighten out, but not without another molt. Mantids will usually still be alive after a bad molt and if capable of feeding, they will survive until they get stuck in the subsequent molt. Some hobbyists feel the proper way to euthanize a mantis severely mangled from a molt is to place it in the freezer. I have a hungry orange-head cockroach *Eublaberus posticus* colony that is far closer to the order of instantaneous. An acceptable methodology for euthanizing severely damaged animals is something the hobbyist should consider before keeping mantids.

Slipped molts are very rare but I have seen them on a few imports, including a large batch of orchid mantids in the late 1990s. Disturbances or lack of moisture can prevent a molt from getting past the initial stage and the two exoskeletons are visibly separated by a fraction of an inch. The animal still functions for weeks or months because the molt only slipped instead of failed but the next molt cannot occur and affected mantids will never mature.

Desiccation or excessive humidity can cause mantids to fall over dead but dead

Ghost mantis nymph with regrown raptorial still nonfunctional after two molts.

always the result of improper conditions. Mystery deaths can occur but are rare and primarily restricted to the first few instars.

Sometimes every hatching nymph from a large ootheca will fall over dead within a day of hatching. This is rare or unheard of for most species, but was common for *Hierodula grandis* (the real reason this species did not last long in culture) and is not terribly rare for *Tenodera sinensis*. This is usually caused by oothecae getting just enough moisture for hatching but not enough for the nymphs to emerge healthy. Unfortunately it is not always easy to prevent and spraying them down after hatching will not help. Excessive spraying can kill hatchlings that were not going to die.

Ghost mantis nymph with left, back walking leg regrown after one molt. The leg is smaller but fully functional.

nymphs are usually due to obvious causes. For most species, especially after the first molt or two, deaths should be extremely rare. Bad molts are most often caused by dehydration but excessive dryness does not wait for a molt to cause death. Excessive humidity that causes death is tied in with lack of ventilation and molding items in the cage, so the actual cause of death is suffocation through lack of oxygen or mold growth in the trachea system. Large, fat animals are far more susceptible to suffocation and less susceptible to dehydration. Small nymphs are (as surface area to mass would suggest) the reverse. If animals are dying, they are not getting the proper amount of air, food, or water. Some species require a more narrow set of conditions but deaths are nearly

KEEPING THE PRAYING MANTIS

A very old female.

Brown spots on the eyes often display no external damage when not caused by caging. The cause is unknown. This is a freshly caught, wild, female *Tenodera sinensis*.

Three molts after the loss of a raptorial leg and it just now becomes useful, though undersized.

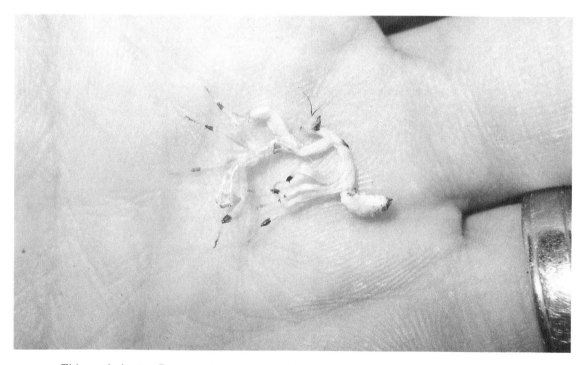

This early instar *Parymenopus* did not molt successfully. It is alive but effectively dead since the legs are permanently deformed and long-term survival impossible.

It is very difficult to overfeed nymphs though extremely fat nymphs are more likely to sustain mortal injuries if they fall in the cage or are dropped accidentally. Adult females often have incredible appetites and will eat until the abdomen looks nearly ready to burst. This can result in large oothecae but also damage and death. Overfed females placed in large cages or small rooms with sharp branches or other objects can end up with sliced open abdomens. Of course dropping an obese female on a hard surface would be little different from a water balloon. A less obvious problem is obese females more readily suffocate if the humidity is too high and the ventilation inadequate.

If an adult female is accidentally dropped on a hard surface, the wound can be sealed with superglue and paper to prevent flies from laying eggs in the wound. If she survives she may still make an ootheca. Nymphs are much lighter and less prone to damage but can sustain similar injury. They should be allowed to heal on their own as glue risks a bad molt which is more of a concern prior to maturity.

Lethargic behavior usually signifies impending death unless the animal is about to molt. There are various causes including old age but the most common culprit is probably waiting too long in between feeding or watering. This should not be confused with a stationary mantis, as many mantids sit still for long periods of time.

Loss of tarsi on nymphs or young adult empusids has been blamed on the use of metal screen, as the protarsi can get stuck and break off. Metal screening is easily covered in cloth or nylon window screen on the inside surface of the cage. There is no need to tear out the screen, and crickets chew right through cloth or nylon, so the prey items being used should be considered if a different screen is used. Tarsi loss may be more related to rapid drying of the extremities that results in brittleness. Brittle tarsi can be caused by low humidity or the rapid drying that results from the convection heat of incandescent bulbs. I used metal screening with *Gongylus* and did not experience this commonly reported problem, but I also did not use incandescent bulbs.

Brown eye spots caused by caging.

Misshapen wings occur during the final molt and are most often caused by insufficient humidity. Inadequate nutrition may be a secondary cause while the uncommon individual will try to hold on with the back legs as the wings are expanding resulting in bent wings. Undersized caging can be another cause but that should go without saying. Ruffled and defective wings usually have no effect on longevity unless other parts of the mantis are deformed.

Rhombodera stalii female with misshapen wings.

Ootheca wasps are normally tiny and have enlarged rear legs. Wasps may only affect a few eggs. An emergence hole is visible lower left.

A "floppy abdomen" may occur in some mantis species that do not naturally bend the abdomen but constantly hang upside-down in a captive habitat. This individual was able to heal after the cage was kept on its side.

Brown or black spots are often seen on the surface of the compound eyes. They are often the visible result of rubbing against the side of the cage and continued damage may be prevented by changing the type of perch or moving the animal to a larger or taller cage. If the animal is immature, the spots can heal partly following each molt, but for adults the habitat change will only help to reduce further damage. The spots can also result from contact made with objects during the molt and may display no surface scarring. It is also possible to find wild specimens of aging adults with a similar looking problem, such as the female *Tenodera sinensis* pictured, and in these cases it seems dried blood pools in the eyes as the animal ages. There may be causes beyond physical damage and old age, but even when most of the eye is darkened it rarely affects prey capture under captive conditions.

As mentioned earlier, old specimens may show signs of aging before they die. Parts of the body can fall off or turn black. Old adults may spit black fluid resembling tar. Geriatric females often can no longer form good oothecae while aging males lose interest in females. There is no chance of healing the geriatric mantis but longevity, and possibly comfort, can be enhanced. Old mantids often require hand feeding or the creatures starve. Conditions or perches should be offered which prevent them from constantly trying in vain to reposition themselves. Some hobbyists go so far as to glue on artificial limbs as the old ones disintegrate.

There is little to nothing that can be done with parasitized wild animals other than thoroughly exterminating any parasites that appear. Wild mantids are sometimes infected with tiny nematodes (Phylum Nematoda) or gigantic horsehair worms (Phylum Nematomorpha) but even when imports were common I never saw any of these. By the time you get the mantis, it has already had most of the internal organs eaten, so treating for worms will not do much good. Wild oothecae in the U.S., especially *Stagmomantis* (pers. obs.) are sometimes infested with parasitic wasps (Order Hymenoptera: Superfamily Chalcidoidea). All you can do is confine and kill emerging wasps and hope some eggs have been spared. There are also parasitic flies that kill and emerge from wild-caught mantids. Afflicted mantids are often immature late in the year so abnormally late development is a good indicator of a parasitized animal in the wild (Rick Trone pers. comm.).

Galinthias amoena (Family Hymenopodidae) cleaning antenna, not eating it.
© Kenneth Tinnesen

MATING

Mating mantids and ensuring fertile eggs is the most important part of rearing mantids. If the ova do not develop there is no need to go further. For the majority of species this is the single most difficult aspect, though even easily mated species can be lost with bad luck and lack of foresight.

We will begin the mating discussion with a few exceptions. Our southeastern Brunner's mantis *Brunneria borealis* is unique among the world's mantis fauna, as it reproduces only through parthenogenesis. No males exist. Unfertilized eggs develop into females that are essentially clones of the mother. A few members of other genera including *Miomantis* and *Parasphendale* can develop this way, but only a few eggs, if any, hatch, and survival is usually very low. I kept generations of two species of both genera and never saw an infertile ootheca hatch or hatch all females, so it is more an exception than a rule. Parthenogenesis has been reported in a dozen other genera, but is extremely rare and, so far, not a single long-term hobbyist stock has been established in this way. The final exception to the primacy of mating is a wild-caught adult female. Even if a wild-caught female appears to have recently reached maturity, her fertility is more certain than if you observe a pair mate in captivity. In the case of mating, nature is very thorough.

The first step in mating mantids is the determination of gender. Adult male and female mantids are sexually dimorphic. This dimorphism is difficult to see in some species, while blatantly obvious in others. To identify adult gender, look at the tip of the adult's abdomen.

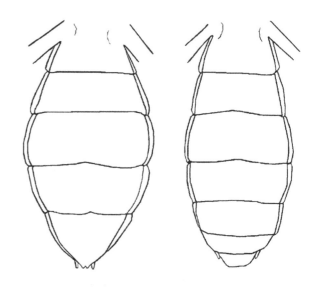

Abdomenal segmentation. Female (left) and male (right) after 5th instar *Sphodromantis lineola*.

The male's abdomen ends in a small, often unequal segment, while the female's terminal segment is the largest and ends in a notch. The male's cerci are commonly much larger. He has a thin abdomen and often trades in some of his camouflage for larger or longer wings. The female's tegmina can be larger but not relative to the body, and only so much as they bulge out to cover a huge abdomen. In a number of

Phyllovates chlorophaea.

Hierodula membranacea.

mantids, such as members of the genera *Stagmomantis* and *Thesprotia*, the male has long wings while the female's wings are short or altogether absent. The adult males of most species are excellent fliers. Males are much lighter bodied and more skittish than females. In many species the male has much longer and thicker antennae, while in the family Empusidae the male has feathery antennae. The three simple eyes on top of the head are comparatively large on males.

There is a way to definitively determine the sex of mantids that appear similar—especially nymphs with greatly muted sexual dimorphism. The terminal abdominal sternites (last few abdominal segments on the underside) appear different. A look at the terminal segment may be enough. On the male, the last visible segment is tiny compared to the other segments. The last visible segment of the female's abdomen is comparatively large, often the largest or widest segment. Sexing of each nymph using the last visible abdominal segment criteria may require a magnifying glass, but is one hundred percent accurate unless they are in the early instars. In a number of species small nymphs appear to be all males until the fifth instar because both genders have the same number of segments. Segments become covered over by other segments as the genders develop. On older animals, segments counted from the abdomen display as five on the female and seven on the male. The only problem with this method in regards to older nymphs is that on a few species the segmentation is difficult to see due to the coloration and shape of the segments.

It is necessary to identify the male and female early on so that maturation can be timed for mating by adapting husbandry. Most males grow to adulthood and die much more quickly than females. In nature, males who mature late are likely to see no action. Some males require fewer molts to reach adulthood than females. Even in species with the same number of molts, the males are much less massive and do not require as much time between each molt, especially if fed the same amount. An adult male flower mantis of a species like *Hymenopus coronatus* or *Theopropus elegans* can reach adulthood months before the female, but seldom lives more than a few months after maturation. Even if the male is still alive when the female molts to adulthood, he may need to live a few weeks or more before she is ready to mate. Old males are less viable even if they still have a desire to mate. Varying the temperature and amount of feeding affects how long a mantis lives and how soon it molts to adulthood. The female nymph can be kept in a warmer spot and fed heavily, while the smaller male can be kept on a cooler, bottom shelf and fed smaller or fewer prey. If sexes are determined early on, it can be rather easy to synchronize maturity with a little practice. Underfeeding and cool temperatures can result in death even if they are not terribly extreme. Males can be kept and fed normally while females are "power fed" with as much and as large of prey as possible. There is a good chance maturation is more closely synchronized in nature because females can eat as much as they want. Males are likely to mature earlier in the wild but huge disparities are artifacts of captivity. It is not always males that mature early, as *Pseudovates arizonae* females can mature many months before males from the same ootheca. If young can be acquired from oothecae hatched weeks or months apart, it may not be necessary to keep the two sexes differently, although males from the first ootheca and females from the last are unlikely to have mates.

Another reason to check the gender during the early instars is to help determine how many

KEEPING THE PRAYING MANTIS

Headless males are not terribly rare even with less aggressive species like *P. chlorophaea*. More voracious species usually finish the whole thing.

Rhombodera stalii, a mean female. He was the lucky one. The first three did not make it aboard.

nymphs will be kept. It is important to get the sex ratio to the desired number before getting rid of any and it is usually better to keep more males than females. The number of adult mantids kept or available is a huge factor in maintaining multiple generations. For some species like *Hierodula membranacea* and *Miomantis paykullii*, it is possible to get away with keeping one female and one male for a number of generations, but sooner or later the stock will be lost. In general it is best to keep at least six males and six females to ensure the next generation. However, successful long-term breeders of more difficult to mate species like *Hymenopus coronatus* keep many dozens of each sex from multiple oothecae and different pairs. It is not possible to keep too many for mating, but it is easy to keep too many to properly feed and care for.

A pervasive myth regarding mantis mating is "the male has a specialized inhibitory nerve in the head to prevent him from mating and the female liberates his libido by chewing off the offending head" (reviewed in Prete & Wolfe 1992; Prete et al. 1999). This myth arose from a misinterpretation of some interesting experiments K. D. Roeder performed in 1935. He observed successful matings and coordinated movements in decapitated mantids and concluded that this feature was an evolutionary response of male mantids to predation risk from females. However, he never stated that decapitation was a necessary component of mantid courtship and mating. Nevertheless, it was this interpretation that persisted and even found its way into entomology textbooks. A headless male cannot chase down a female and even males *with* heads can have trouble getting on the right way. In many species males are never or rarely killed. There is a grain of truth to the myth as males often seem to make attempts at copulation when the head is

Tarachodula pantherina mating. © K. Tinnesen

removed, but a variety of unusual behaviors specific to the abdominal nerve center are also known to occur. Since he is most likely to lose his head during mating, if he went into convulsions and was not finished, he would have lost his head in vain, but this also happens. Some males have no difficulty mating with their head gone, but females have also laid oothecae in the same condition. Yet a headless female laying an ootheca is not touted as an adaptation. Vertebrates often defecate when beheaded but it is not generally considered an adaptation to relieve constipation. Even if placed immediately on a receptive female that does not immediately grab and eat his remains, the chances of successful copulation are very limited. It is not a given, for if the male is not

In certain species or individuals the spermatophore transferred to the female remains visible after mating. *Tarachodula pantherina*. © Kenneth Tinnesen

Close-up of recently mated female with spermatophore visible. © Kenneth Tinnesen

Sexual dimorphism. In some species the male is tiny compared to his counterpart, *Hymenopus* adult pair.
© Kenneth Tinnesen

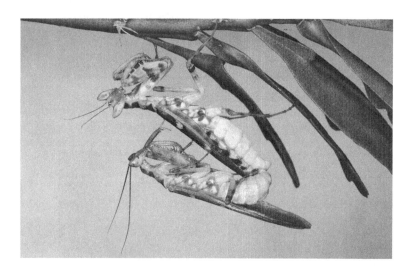

Creobroter sp. mating.
© Tammy Wolfe

Polyspilota aeruginosa, neither of these males appear to be having much success.
© Tom Larsen

already responding to cues to mate when his head goes missing, he may just twitch. I used to chuckle when I imagined mantis breeders sitting around with a pair of scissors ready to mate a prized pair, but only because I did not think anyone really would.

The adult appetite plays an important role in mating, but not so important as the myth suggests. As nymphs, males and females are both voracious feeders, but after adulthood the male eats very little. If an adult male were to continue feeding at the same rate after maturity, he would not be very agile and would be too heavy to search out a female. He would also have a difficult time escaping from the hungry female. The male does not eat the female because she is usually much stronger, but mostly because she is too big a meal for his delicate appetite. The female, on the other hand, has an unquenchable appetite after reaching adulthood. Females sometimes do eat males, but this is the exception rather than the rule in nature. In captivity males often get eaten if mating is tried in tiny cages where they cannot escape the female's great appetite. The female should be heavily fed to help keep her hunger from overwhelming her desire to mate. Remember that in nature the male has multiple escape routes. In captivity, if the male is placed with a hungry mate and not given numerous places to hide and escape, he will be eaten.

In captivity most male mantids are ready to mate two weeks after becoming adults. Only in some of the tiniest boxer species is the male ready in a week or less, since he must be fully sclerotized. Old males (age being relative to the species, but generally over two months) can lose interest in mating or be rejected by the females. Usually after both male and female have been mature for two to three weeks and the female has been fattened up, they are placed together for mating. Some breeders will wait till the female displays "calling" behavior, as she arches the abdomen to broadcast pheromones.

It is difficult to determine how soon mantids mate in nature. Since mated females are often no longer receptive, it would seem mating would take place shortly after the final molt, possibly while the exoskeleton is still soft. Males certainly require time for the exoskeleton to thoroughly harden, but a female could mate immediately after a molt. Although previously documented, only recently has anyone realized mature male Blattodea of a number of species mate immediately, while the female's exoskeleton is still soft and pale (McMonigle 2012a). Teneral mating in captivity poses a problem for mantids as the male can overpower and eat the female at this time. This could be a good part of the reason the male's appetite is so wimpy. The earliest documented time for mating following a molt is the second day for a female *Gongylus gongyloides* (report by Yen Saw). The male was three weeks old. I have acquired wild females that were a little soft and thin and could not have been adult for long, but their oothecae were always fertile. In captivity it is possible to watch a pair mate and still get infertile eggs.

Pheromones have a different effect in captivity than in the wild. In nature, adult mantids are dispersed throughout a meadow, garden, forest, etc. Females tend to stay in a small area while the males fly around looking for mates. Males are attracted by species-specific pheromones the females broadcast. Males do not try to mate with large female nymphs that do not yet produce these pheromones. When the pair is in close proximity visual cues take over (Lasebny & McMonigle 2001; Gemeno et al. 2005; Gemeno & Claramunt 2006; Maxwell et al. 2010a, b). In some species there is a ritualistic dance, while in others the male simply jumps on the female's back while keeping a watchful eye on her dangerous forelegs. In captivity both male and female live in a relatively small area and usually in the same room. The airflow in a room is stagnant compared to that of the outdoors and pheromones from mature females build up to thousands of times the levels found in nature. Adult males of some species could become desensitized to the pheromones and lose their desire to mate. Males may still go through the routine of mating (possibly to avoid being eaten) but without fertilization. Desensitized males often do not even bother with the ritualistic dance and have no desire to jump on the female. Outside ventilation or relocation of males to another room may be necessary to remedy this situation (Lasebny & McMonigle 2001).

A variety of variables affect the interest in mating. In some species it seems males show little interest in mating if kept at 75° F or below. Empusids and *Phyllovates* may show little interest in the female unless they are able to raise their internal temperature with a basking light or the ambient temperature is above 80° F. Competition also seems to play a role as mating often commences much more quickly if two or more males are placed in the mating cage. Background noise and vibrations may play a small part. As of yet no experiments have been run to determine the effects of rainforest recordings or Barry White music. Time of day and lighting are other factors. One manticulturist reports that one male *Sphodromantis* would only mate with the females when the lights were turned off (Gale 1992).

Hand pairing is the safest method for mating mantids, but most species do not respond well to being handled. This method has been used to successfully mate a handful of different species without the male getting killed. The male is placed on a horizontal surface (such as a table or floor) or a vertical surface (such as a rough wall or screen cage) so he can clearly see the female. The male is set on the surface and watched for signs he has calmed down after being handled. This can take great patience when he is restless and constantly runs or flies away. After he has settled, the female is slowly placed near him, but a few inches above or past him facing the same direction. It is important she is placed facing away from him, not toward him. If the two remain still a gentle tap of puff of air may get her to walk away slowly. The male should recognize her movement and begin a cautious approach. He may initiate a courtship ritual by waving his abdomen and forearms or lifting his wings to expose a brightly colored abdomen. If the female walks too fast and he does not appear to follow it will be necessary to try again. When the male jumps on her back, a piece of cardboard or your hand should be ready to be placed in front of the female if she attempts to grab him as he is positioning himself.

After they are mating, and the male's abdomen is firmly attached to the female's, she can be given a prey item to chew on. They should be left alone to mate for at least an hour. If all goes well, the male will be on her back, unattached, or a distance from her. The male can then be used again with other females. It is always easy to get the male on the female's back, but when placed there, he usually sits there and does not consider mating or she turns around and takes a big bite out of him too quickly to be stopped. Losses are not restricted to a bitten male. A fat female can be dropped during handling or the male's flying capabilities can be underestimated. Not every keeper has the knack for hand mating and many mantids just will not go for it. This only consistently works well with a few easy species that are already rather easy to breed.

Popa spurca pair. © Andrew Nisip

Glass terraria or plastic boxes are commonly used for mating introductions. The cage is heavily planted or filled with sticks and branches to offer some avenues for escape, and a pair is placed inside, one on each side. A handful of crickets, cockroaches, or other active food should be put in the cage to provide food and distraction for the female. It is important to watch for the first ten minutes to monitor for initial aggression, but few keepers have the ability or desire to stay perfectly still for the next 2-8 hours. Within a few hours to a day the pair should be seen mating. Small (ten-gallon), decorated terraria work great for ghosts and boxers. Large, aggressive mantids like *Sphodromantis lineola* and *Tenodera*

Pseudocreobotra ocellata.

Stagmomantis carolina mating on a *Ficus*.

angustipennis can be mated in similar set-ups, but the terrarium should be at least a thirty-gallon high to give the male a good chance of escaping the female's voracious appetite. Females often eat every male without mating if there are no adequate hiding places. When I kept *D. desiccata* for seven generations, I was only able to get them to mate while they hung below the screen lid of a 70-gallon aquarium with very little else in the cage. Without this large, horizontal surface, males were very rarely eaten but also would never mate. This will also work with some of the big empusids like *Gongylus*, but with most species the large horizontal surface give the female easy access to eat the male. As soon as mating is over the male should be returned to his cage. The female can be mated again with a different male, if possible, to ensure fertility of the eggs. If she mates readily the second time, the first one was probably not successful. Always remove the male, mated or not, after two days. He can be used again after she makes an ootheca, or immediately with another female.

It is helpful to recognize an unusually aggressive female and not waste males on her. Most females of a given species have a similar disposition towards males, but a few will eat every male they are given access to. A different female should be used and males given to this female only after all the rest have mated. I have run into such females a few times across various genera and the first few times I recognized the problem too late. It is often the first female to mature.

I have hand-paired a few species, and used terraria for certain species, but the vast majority of successfully matings have employed a slightly different method—the "flight cage."

I have a large screen cage that looks like a small, screen, phone booth. It is approximately 3' x 3' by 6' tall and contains a couple of large branches. It is made from inexpensive wood and the screen is just stapled and glued on so it could not have been very expensive to make (I picked it up for free). Mantids are placed in the cage out of each other's direct sight and the male is allowed to take as many hours as he wants to make an approach, often on the vertical sides of the cage. I have observed a few approaches in nature and have seen males ten feet away move less than a foot an hour. I have

Mating purple boxer mantids. © Henry Kohler

checked on progress for hundreds of matings in the flight cage and the approach is similar. They do not want to be eaten and so watch her for signals, or the pair simply stares at each other as he imperceptibly inches forward. Observation should be done from as far away as possible since mantids can see humans—movements closer than ten feet may disrupt or discourage mating. Using this method, males do get eaten, but I can keep an eye on the pair from a distance while I work on something else. Usually the pair is found coupled 3-8 hours after introduction. Larger "flight cages" can be better and worse. A large greenhouse full of plants ought to be perfect for most species, but the animals can become permanently lost and most hobbyists do not have greenhouses. Another option is to give the pair a bathroom to mate in, but even small rooms can offer unexpected opportunities for the female to become damaged or lost, and observation for problems cannot occur because the door has to be opened for entry.

Specimens with deformed wings are difficult or impossible to mate. Very slight deformities are not a problem, but only if they do not affect the overall shape and function or are on short-winged species. The wings are used to recognize mates and send signals. Males rarely look twice at a female with deformed wings and even if they jump on the female they are unable to maneuver the abdomen around her deformed wings. If there is only one female and she has ruffled, deformed wings, the only viable option may be to cut her wings off a little bit out from the base to offer an opportunity for the male's abdomen to reach around. It generally will not affect her lifespan, but males may be a bit put off. Males with deformed wings are easier to mate than females, but are likely to be eaten, forcibly rejected, or ignored. Deformed wings can be a sign of other health problems and even if mating can be accomplished, oothecae may be hopelessly deformed.

Most male mantids hang on to the females long after fertilization is over. If coupling only lasts a few minutes it is unlikely any transfer has occurred. Transfer of the spermatophore occurs within thirty minutes independent of how long the pair stays coupled. It has been theorized the male stays coupled with the female in order to prevent nearby rivals from coming in and displacing his sperm (Prokop & Vaclav 2005; and many others), but males simply do not stay long enough. A few hours or less of mate guarding cannot possibly provide significant protection. Since the male is safest behind the female, it may be simply that he is waiting for a good opportunity to escape without being eaten. If adults are still coupled after a few hours and the hobbyist cannot check in on them later, the pair can be carefully separated. Holding the male and female by the thorax will give them the impetus to separate without any need to pull them apart—pulling them apart can cause damage which may prevent adults from mating or properly forming oothecae in the future.

Rhombodera stalii pair. Most females will not eat the male with limited effort on the part of the keeper.

The last, and possibly least, aspect of mating covered here is the idea of inbreeding and declining fertility. "I have noticed a reduction of vigor throughout successive generations of mantids, resulting in the eventual loss of all cultures" (Gale 1997). Improved husbandry knowledge has moved this topic from a wildfire of mantis breeding discussion to a whisper. Mantids are difficult to rear so there has been a strong tendency to use inbreeding as the cause for every mating failure even when such failures occur in just one or two generations. The easiest to keep species were no longer attributed "inbreeding" problems after a dozen or more consecutive generations, but if there were a genetic weakness specific to the Mantodea (no such weakness has ever been discussed for cockroach cultures, some of which have been inbred for forty years) it ought to effect *Sphodromantis, Hierodula, Miomantis,* and *Phyllocrania* as much as *Hymenopus* or *Gongylus*. I remember how excited I and a few friends were to discover by the end of the 1990s that we were keeping species for five, six, and eight consecutive generations when we had been told two or three was the maximum. Some of the more difficult species do seem notably easier to mate the first generation or two, but a good portion may well be related to the excitement of attention paid to a new species. Breeding only brother to sister generation after generation cannot be good for genetic diversity or vigor and can magnify genetic disorders, so populations should not be kept too small and defects should be weeded out. However, blaming the loss of a stock on declining fertility due to inbreeding is nearly always an excuse for inadequate number of specimens retained or improper husbandry.

OOTHECAE

Every mantis species creates protective structures for the ova called oothecae. These come in many shapes and sizes and contain from less than a dozen to hundreds of eggs. The female attaches the base to a surface and lays eggs in rows two to eight across and many times as long. While the eggs are laid, the female exudes a protective casing around them. Oothecae are designed to protect the ova from environmental dangers, but it can be difficult to provide a captive environment that is perfect for gestation.

As the eggs are laid, a proteinaceous fluid is secreted around them. Appendages at the rear of the female's abdomen form the substance into a thick foam, as in *Tenodera*, or a thin casing, as in *Brunneria*. Some are constructed with huge air spaces between the inner and outer layers, while others are attached to thin strips of casing material and will be dropped to hang when finished. It is amazing to watch construction of some of the more unusual oothecae. When completed, the structure is extremely soft and starts out green, blue, or white. Over a few hours to a few days, the foam hardens and changes to brown or tan. The vast majority are brown, but a few species create oothecae that are green, red, purple, or yellow.

The casing protects the eggs from desiccation, predation, and temperature extremes—exposed eggs have zero chance of survival. Ova of temperate and arid species need to survive exposure to very hostile and variable environments. Their oothecae may be laid in protected spots and often consist mostly of protective foam. Oothecae cannot insulate from long periods of cold but do protect the eggs from rapid temperature changes. External oothecal walls of tropical and subtropical species are often thinner; their oothecae consist more of eggs than foam. A thick outer casing may also offer some tropical species protection from chalcid wasps that commonly emerge from wild-collected oothecae. Keeping long wasp ovipositors at bay is a possible reason for the massive air space between the inner and outer layer of the unusual *Hoplocorypha* oothecae. *Acanthops* and related genera form oothecae to hang at the end of a thin string of the protein casing—obvious protection from predation. Cryptic colors and shapes likely protect some oothecae from being eaten, though most are pretty easy to see from a distance. Thomann (2002) explains that *Stagmomantis carolina* oothecae look very much like galls that are common on the same branches they are formed. The surface of oothecae from species found in desert areas can be soft and porous to absorb moisture, while those from rainforests are often smooth and water repellent (Larsen 2007). The shell-shaped projections on *Gongylus* and *Empusa* oothecae are thought to

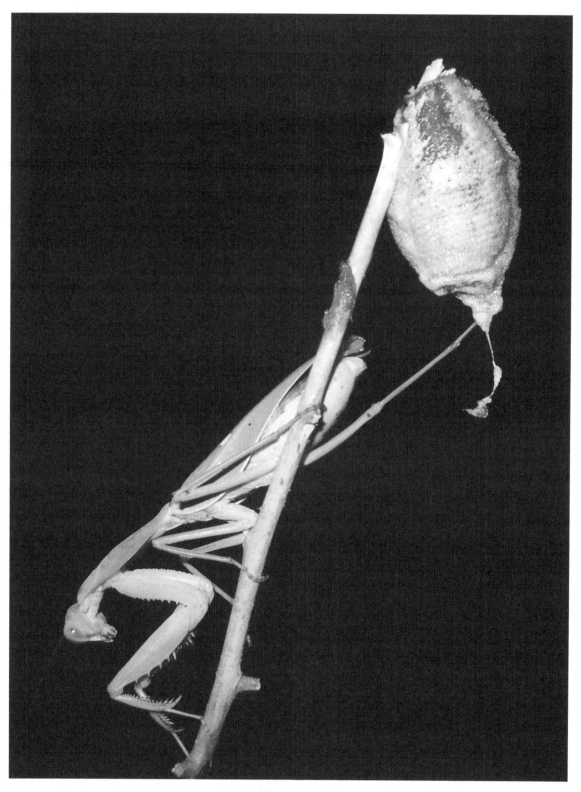

Rhombodera stalii with recently completed ootheca.

channel away heat (Larsen 2002), while the little flags of excess material off one side are hard to imagine a purpose for. The odd, finger-like projection on the *Phyllocrania* ootheca is likewise difficult to explain. Many of the shapes and patterns found on different oothecae probably serve no purpose, but allow many to be identified down to the genus or species.

Even the best laid plans can go wrong. Wild-collected oothecae can be dead, hatched, dried out, or parasitized. *Podagrion pachymerum* and *Mantibaria seefelderiana* are parasites on the eggs of *Mantis religiosa* in Europe. The latter wasp is also an ecto-parasite. The female cuts off her own wings after finding a female mantis and drinks her host's blood from between the base of the tegmina while waiting patiently. She waits for the mantis to form an ootheca so she can lay the eggs inside before the mantis can finish the thickly protective, foamy exterior (Hutchins 1966; Berg et al.

This is an infertile, dried out, *Polyspilota aeruginosa* ootheca.

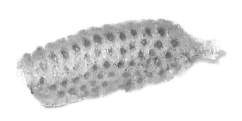

The tips of the cylindrical eggs are visible through the bottom of this *Litaneutria minor* ootheca.

2011). Various parasitic wasps attack the oothecae of different species. In the United States, *Podagrion mantis* and *P. crassiclava* are known to parasitize *Stagmomantis* oothecae (Berg et al. 2011). Although the wasps are little bigger than fruit flies and display no obvious defenses, small mantids that hatch from the same ootheca never seem to eat them. I have tried leaving in wasps so nymphs could avenge their lost siblings, but they eat each other before they eat the wasps.

The maximum number of oothecae a female produces during her life varies from species to species. One female can make from two to more than twenty oothecae. Some can make the next one after just a few hours (*Oligonicella scudderi*) or they may require a month in between (some *Sphodromantis* spp.). A few weeks in between is common. Although females only need to mate once to produce many fertile oothecae, fertility often decreases within a few months. After a few oothecae are produced, it is helpful to re-mate the female; otherwise, the final one or two may not hatch or may produce a very small number of hatchlings. Infertility can delay ootheca construction, but has little effect on ootheca formation. Unfortunately, infertile oothecae commonly look identical to fertile ones.

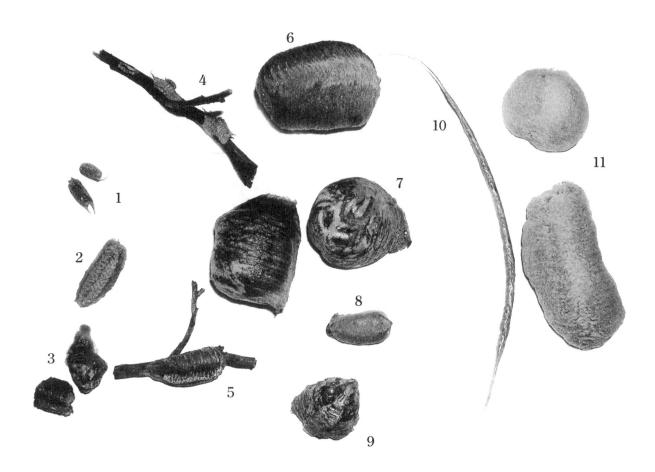

(Above) 1. *Otomantis scutigera* (2). 2. *Pseudogalepsus nigricoxa*. 3. *Phyllovates/ Pseudovates* (2). 4. *Oligonicella scudderi* (4). 5. *Stagmomantis carolina*. 6. *Sphodromantis lineola*. 7. *Hierodula membranacea* (2). 8. *Popa spurca*. 9. *Deroplatys desiccata*. 10. Acanthopidae. 11. *Parasphendale affinis* (2).

(Opposite) 12. *Hierodula patellifera*. 13. Acanthopidae. 14. *Paraphendale affinis*. 15. *Acanthops falcata*. 16. *Hierodula membranacea*. 17. *Phyllocrania paradoxa* (3) 18. *Sibylla pretiosa*. 19. *Oligonicella scudderi* (4). 20. *Popa spurca*. 21. *Blepharopsis mendica*. 22. *Pseudgalepsus nigricoxa*. 23. *Oxyopsis gracilis*. 24. *Gongylus gongylodes*. 25. *Gonatista grisea*. 26. *Stagmomantis carolina*. 27. *Taumantis sigiana*. 28. *Phyllovates chlorophaea* (3). 29. *Otomantis scutigera* (2).

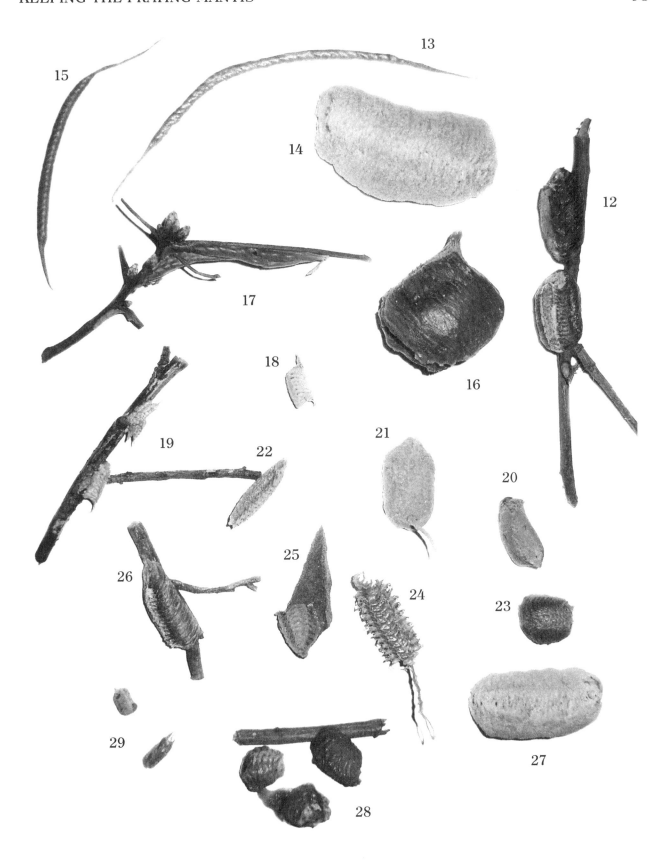

Mated females should be fed as much as possible so oothecae are large and healthy. Volume of prey offered to the female affects shape, number, size, and health of oothecae. The quality of the food is also very important. A nutritious diet of *Nauphoeta cinerea* roaches along with some mealworms and crickets produces impressive oothecae, while a diet such as house flies and waxworms can produce inferior oothecae, with the exception of empusids. The oothecae of poorly fed females are not only small, but are often half-formed and oddly shaped. Not only do small eggcases contain fewer eggs, they also tend to have a lower hatch rate of the eggs present. Poor feeding can cause deformed oothecae that expose part or all of the eggs to desiccation. With severely limited feeding, a female may never even form an ootheca.

The type of egg-laying sites can be just as important as the female's diet. Depending on species, females should be provided with twigs and branches of varying diameters, or flat surfaces like a piece of tree bark. (A flat surface will change the usual shape of the ootheca.) Not only do different species form oothecae on different kinds of surfaces, but some individuals of the same species are more discriminating. Some females hold the eggs in until they find a suitable place to put them. Picky females can become egg bound and die if they do not find a suitable site. Or, a female may eventually give up and just dump the ootheca contents on the floor of the cage and none of the eggs will survive. Even when many different sites are provided, she may decide to make it right on the screen.

Access or gripping surfaces are important because if the female falls during formation of the ootheca, she will not come back and complete it. The remainder will be formed elsewhere. She does not just form a second smaller ootheca with the remains, but rather the other half of

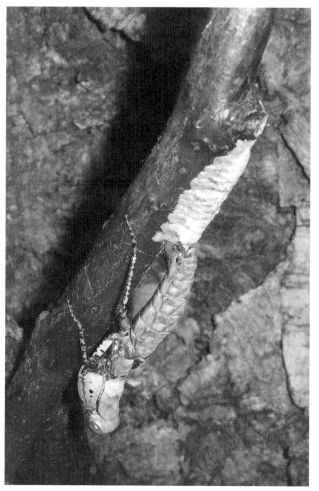

Tarachodula pantherina forming ootheca.
© Tom Larsen

the unfinished one. This is not terribly noticeable if oothecae look similar on the top and bottom, as the base will just be wider.

It is possible to induce a picky female. If a mated female has a greatly expanded abdomen and seems overdue, misting a few times a day to increase the humidity can work. If higher humidity does not help, she can be moved to a larger cage, placed in a planted terrarium with multiple laying areas, or "free-ranged" on a large houseplant (Lasebny & McMonigle 2001). Sometimes she will find a good spot in half an hour. If, after a few days and a few environments, she still refuses to lay, you can try restricting feeding or mating her again. Allowing her to process the food already in the abdomen may help since extreme obesity can be a culprit. If none of these work, there is probably nothing that can be done.

Oothecae should be removed from the adult's cage as soon as the foam hardens. Immediately remove prey animals, if present, because they are especially attracted to the delicious, soft, fresh oothecae. Old hardened oothecae are far less tempting. Prey offered to the female may not destroy a dry ootheca, but are likely to eat holes in it. The female does not feed on her egg mass like many arachnids, but she will eat her young when they hatch in the cage. In stark contrast to the normal experience of mantis breeders and familiar mantis behavior, K. & R. Preston-Mafham (2005) report a tarachodid female *Oxyophthalmellus somalicus* in Kenya was observed guarding her nymphs. Another reason for removal is oothecae of temperate species may need to spend some quality time in the refrigerator.

Never attempt removal before the ootheca fully hardens, which can take 24 hrs or more. For the first hour or so it is gooey and can be utterly destroyed if handled. The more intricate the eggcase, the worse the damage. After a few hours it is more solid but is still soft and emergence channels can be easily damaged. Mantids that form oothecae on branches allow for easy removal because the branch or twig can be taken out of the cage immediately. Removal of oothecae attached to larger decorations or the cage poses some risk as the ootheca can be damaged though a good tug usually breaks it off clean once hardened. If attached to a screen vent, a new, sharp razor blade may be necessary to carefully cut it off. Be careful. If any portion of the back of the eggcase remains on the surface some of the eggs can be exposed and damaged. Although eggs in a damaged section usually die, the unexposed section should still

Ootheca formation.

hatch. The female may be moved from cage to cage and the old container kept, in which case there is no need to relocate the ootheca.

After removal, the ootheca can be re-attached (in the same orientation) to another surface. The orientation is important because the eggs face one direction and oothecae are equipped with a line of weakness or "hatches" through which the nymphs exit (Farb 1962). Opposite the point of attachment, or opposite the back if a hanging ootheca, there is a hatching zone made up of predetermined lines of weakness that often runs the length of one face of the ootheca. Hatchlings will emerge head first from here. On many oothecae, the hatching zone is very conspicuous because it is a different color or texture. This often makes it easy to tell the 'front' or point of emergence. On some oothecae like *Parasphendale* the apertures are not a different color but the shape of the ootheca makes it easy to see where the hatching zone is. Some oothecae with a "spine" like *Brunneria borealis* do not have a general hatching zone, but rather lines of weakness that run to a single spine and all nymphs must emerge from the one small point (McMonigle 2004b).

Usually the ootheca will be pinned or glued onto the new surface. It is hard to get the ootheca to sit in the desired position using pins, some eggs can be punctured, and pins can fall when the cage is misted. Super glue (cyanoacrylate) was the popular choice of manticulturists in the 1990s and worked very well. When using super glue, it is applied to the desired surface, the "back" of the ootheca is held against the glued surface and then the exposed edges are sprayed with water. The glue sets up immediately on contact with water. Other glues should be avoided or carefully tested since chemicals in the glues might kill the eggs. The common non-toxic hot glue used in crafts is more commonly used by manticulturists today since it is also popular in cage construction. Oothecae are well insulated but hot glue can cook eggs near the point of attachment and could kill most eggs in a thin, flat ootheca like that of *Litaneutria* or *Pseudogalepsus*. This can be avoided by waiting for the hot glue to cool to the touch (not too cool) before attaching.

Though most keepers hang or glue oothecae, it is not necessary if there is no substrate in the hatching container. As long as the ootheca is not resting on the area of emergence the hatch rate will be related to the care, not the orientation. (Small, lightweight oothecae often

Rhombodera valida hatching. © Henry Kohler

hatch perfectly even when accidentally bumped and found laying on the hatching zone.) Oothecae are simply placed with the back side or place of attachment against the floor of the container. I have hatched out the majority of species in both species lists this way and have learned firsthand the hanging requirement is another myth, albeit a reasonable sounding and intuitive one. Apparently the pressure required to burst the hatching zone is more than enough to counter the effects of gravity on tiny nymphs. The nymphs of some species (e.g., *Idolomantis diabolica*, *Macromantis* spp.) hang from longer than usual threads when hatching, but other species with varying string lengths do not require hanging (even live-bearing cockroach L1s can hang from strings when hatching; see photo). A properly hydrated ootheca will hatch

Rhombodera valida hatching. © Henry Kohler

in any orientation while a dehydrated ootheca will not hatch properly no matter how high it hangs.

Various methods have been used to house and care for oothecae during gestation. A common method involves placing them in a suitable container and misting once or twice a day. Almost any container works well as a hatching container but it should not be overly large and it should have clear sides. Tall 32oz. deli-cups are probably the most commonly used container. The lid or side of the hatching container should have a large screened vent like the mantis cage. Usually an inch of vermiculite, soil, folded paper towel, or coco fiber is placed on the bottom to retain moisture. This type of hatching container should have ventilation and dry out between mistings or eggs can mold and die. Unfortunately they cannot simply be kept dry to avoid mold because eggs inside the best insulated ootheca will wrinkle up and die if kept too dry.

There are many theories on the best method to incubate oothecae. Most experience off and on luck balancing moisture and ventilation affected greatly by seasonal changes, level of attention, individual hobbyist experience, where the incubator is placed in the room, and other variables. The following method is relatively resistant to external variables, requires little to no maintenance, and is well tested:

1. Remove ootheca from the mother's cage about 24 hours after formation. Be sure it is clean and has no wood or any material attached to it that could grow mold.

2. Place ootheca in a 16oz, sealed deli-cup. The container should have a *single*, tiny pinhole for air exchange in case the lid is airtight.

3. Do not ever directly mist or moisten. However, a single drop of water can be placed elsewhere in the cage. If the drop evaporates within three days it can be replaced. If not, remove it with a piece of tissue or paper towel.

4. Date container and wait.

It can take some experience to properly refrigerate oothecae without killing them (refrigerator, not freezer). *Mantis religiosa* usually require a cool period while *Stagmomantis* and *Brunneria* may hatch better, or at least not as intermittently, if cooled for a few months. Another reason for refrigeration is to delay hatching when there are too many oothecae for

This is an 8oz. hatching container showing nymphs did not need to hang. Multiple oothecae are often kept in the same container since they rarely hatch at the same time. They must be moved to a grow-out cage as soon as possible. I toss the ootheca in the growout cage as well in case a few more hatch out.

nymphs to be properly cared for. Temperate oothecae are usually kept in a plastic container with a few pinholes and some paper towel to maintain humidity. Even in the vegetable crisper drawer things can become utterly dry in a refrigerator. Primarily it is important to check on the ootheca every week or two to make sure mold is not growing on the paper towel or container and that the paper towel is lightly damp but not wet. The container should be cleaned and paper towel changed at the very slightest sign of mold. It is very easy to forget there is an ootheca in the refrigerator if a regular schedule is not followed. It is much easier to keep oothecae in the garage or outside for the winter but it is easy to miss the hatching when temperatures rise, later discovering a tiny pile of dead nymphs. Oothecae may be kept refrigerated a month or two longer than they naturally would diapause, but they never seem to hatch out if left much longer.

Most oothecae hatch four to six weeks after they are made, or after being removed from cold temperatures. On the other hand, there are uncommon circumstances where oothecae collected in late winter can hatch in as little as 24 hours of removal from cold temperatures. Do not give up on an ootheca that does not hatch at the expected time. Some oothecae just take longer to hatch or the wrong date may have been written down. The ootheca of *Brunneria borealis* regularly takes six months to hatch if kept at room temperature. Some desert species (*Iris deserti*, *Eremiaphila*) may take longer, and most temperate zone species need this time due to overwintering.

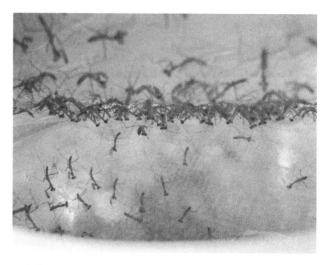

A few *Rhombodera* hatchlings from one ootheca.

Partial hatches are exceptional but not rare. Usually all the young will hatch from an ootheca within a few hours and then it can be discarded even if the number of nymphs is much smaller than expected. Partial hatches are often the result of parts of the ootheca becoming too dry and the affected eggs dying, or the young developing but simply being too dehydrated to emerge. Damage caused by poor removal or prey feeding on oothecae are another cause for a partial hatch. However, a few stragglers may not emerge till the following day, or if the hatch is very small, the majority can hatch the second day. Elevation of humidity is often suggested for small hatches but there is no evidence to suggest the rest were not going to hatch the

Ceratomantis sausurii oothecae (above) and hatchlings. © Kenneth Tinnesen

fertility will help determine what went wrong and what can be improved for future attempts. Dried yellow eggs usually means they were dried out excessively early on and may or may not have been fertile. Gooey eggs with no signs of development means they are infertile (unless the keeper gave up too early and prematurely cut it open). Dead nymphs means they were kept humid enough to develop, but a little too dry to hatch. Observing dissection results will help a keeper hone skills, but can be confusing when a *Phyllocrania* ootheca is opened and some yellow crystallized eggs, some gooey unfertilized, and a few dead nymphs are all found in the same ootheca.

Note the nymphs hanging from strings. This was a huge mass that looked like an improbably large mantis ootheca hatching, but by the time I got the lid peeled back, camera ready, and positioned most of the nymphs had fallen. The hissing cockroach female had chosen to climb to the screen whereas live bearing usually occurs on the surface.

next day anyway. I have observed a number of partial hatches and did not attempt to change the humidity. *Parasphendale* species are well known for this. It is possible for young to hatch out more than 48 hours apart, but with exception of the exceptional *Brunneria borealis*, some *Eremiaphila* species, and *Tenodera angustipennis* in unusual circumstances, it is extremely unlikely.

Eventually it is time to give up on an ootheca. Even after years of experience it is possible failed eggs were improperly cared for or were not fertile despite observed mating. Cutting open failed or partly failed oothecae to check

Remember to eliminate hazards from the hatching container. If there is a small piece of tape or any sticky residue in the container and a few dozen orchid or ghost mantids hatch out, every last one will be stuck to the tape when the hatch is noticed. They must be removed from the tape but even careful removal will result in torn bodies and ripped off legs. Sticky fruits in the hatching container can also trap and kill a surprising number of mantis hatchlings. Another common item in hatching containers that will kill hatchlings is a water droplet. The surface tension can trap and drown one or more hatchlings depending on the size of the droplet and the nymphs.

Unlike others aspects of mantis care, ootheca gestation may still be successful without consistent care. Refrigeration and vented gestation caging will be successful if close attention is paid. Hopefully this section has provided every circumstance a hobbyist is likely to run into when incubating oothecae.

Mantis (unidentified species) with ootheca, from Ecuador. (CC) Geoff Gallice

THE PREYING MANTIS

The number one variable that influences the growth rate, health, and ultimate size of mantids is prey. Without adequate food mantids will grow poorly, take incredible lengths of time to mature, or simply expire. Nymphs may come down and feed on a little banana slice, but prey is not an area that can be finagled and result in success. If it is not possible to provide for the hatchlings' needs, the ootheca should be sent to a good home before it hatches.

Feeding mantis nymphs and adults is similar. The biggest difference in care is the size of the food that can be eaten. The tiny hatchlings only accept small prey such as fruit flies, springtails, newborn roaches, and hatchling firebrats. Larger nymphs and adults eat progressively larger foods. Most species can be fed fruit flies as a singular diet for the first few instars. A few species are more timid and must be started on incredibly tiny springtails or hatchling firebrats. Not all individuals are as aggressive and some may starve rather than try to tackle the larger fruit flies. If kept together, nymphs should be fed as much as possible and more often to minimize cannibalism. It is very important to prepare food cultures ahead of time, but in some cases not too far ahead of time. It can be impossible to maintain enough food to keep up with the demand if there has been no planning.

After the first three weeks or so, nymphs have molted twice and prefer larger prey like cockroaches that do not have to be fed as often and do not necessarily require planning, because established cultures produce at a constant pace. Fruit flies and springtails may still be accepted for a number of molts but are often too small to be noticed by larger species. Small crickets, young cockroaches, adult firebrats, house flies and other medium-sized food should be offered to the growing mantis nymphs. When switching from fruit flies it is important to consider the attitude of the new prey and habitat style. Mealworms, crickets, certain cockroaches, springtails, and firebrats cannot climb smooth surfaces so they will never come into contact with mantids hanging from a tall cage's lid. A ladder may help but can also change where the mantis molts or give access to hungry crickets. Lobster roaches climb, but if they are not eaten right away will hide in the corner. Lastly, very large nymphs and adult females are switched over to adult crickets and cockroaches (such as *Blaberus* and *Blaptica*) since the huge quantities of medium-sized prey they can consume would wipe out feeder cultures.

The amount of work needed to feed mantids is partly determined by the size and type of food animal used. Hardy and less aggressive prey such as roaches and mealworms can be fed in numbers, so the mantis can be provided a food supply that will last for weeks. Potentially dangerous prey such as crickets should be fed

Parymenopus nymph feeding on freshly molted *Nauphoeta* nymph.

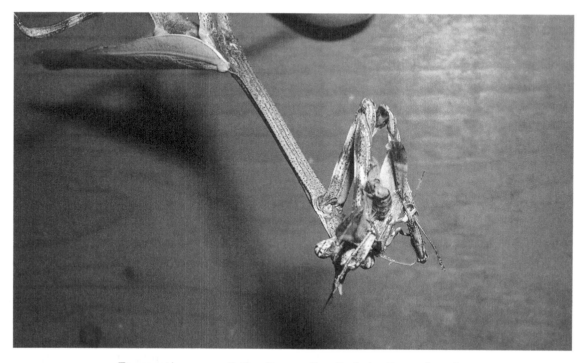

Texas unicorn mantis feeding on *Pyralis farinalis* meal moth.

to the mantis one at a time. Although house flies are not nearly as dangerous as crickets, they are not fed in numbers greater than what the mantis can eat in a day or two. Inside a mantis cage, house flies drop dead (like flies) after a day if they are not fed. As for the size of the prey, it is a good idea to err on the side of caution and not introduce food that is larger than a third of the size of the mantis. As long as large prey is not dangerous, it can be fun to watch one of the more voracious species try to consume an animal much more massive than itself. Larger prey items will fill up the mantis and one or two large roaches may be all the food necessary to get the nymph to its next molt or the female to produce another ootheca.

Mantids do not possess venom to immobilize prey and must have corresponding strength and size to overcome prey. In contrast, assassin bugs immobilize prey with venom, while spiders can use webs to capture and immobilize prey many times larger and stronger than themselves. Contrary to some popular literature, a mantis does not always kill its prey by biting off the head to overpower it. The mantis starts chewing on whatever part is closest to the mouth at the time of capture. Mantis food often stays "alive" during the entire meal and every kicking, biting, and twitching piece must be held tightly as prey is devoured.

There are a number of factors to consider when choosing prey size. Prey that is too large is no better than prey that is too small. Large bulky mantids with thick raptorial forelegs are going to take much larger prey than those with long, thin bodies and skinny forelegs. (A 3/4" boxer will take out a cricket a 2" *Thesprotia* would not consider.) Some species do not look stronger but are just more aggressive feeders. The type of prey is as important as the type of mantis. Many species will take a larger roach than cricket because it does not jump aggressively. However, many species that will not look sideways at a 1/4" cockroach or 1/8" cricket will excitedly massacre a moth with a 1" or better wingspan. Moths taken with zeal can be bigger than it seems the mantis could even hope to attack if it were simply gauging size. Lastly, noting the animal's response is foremost; if the mantis does not show interest it does not matter how much forethought was put into choosing the prey.

Mantids are not always hungry. Nymphs eat well for a few weeks or so and then suddenly lose their appetite. The abdomen usually becomes slimmer as the nymph molts a day to a week after it stops feeding. Molting for subadults is even more obvious because when the final molt (in winged species) is near, the small wing stubs on the thorax become raised up and darken and appear thick and swollen instead of thin and pressed against the body. When the wing buds are swollen it is very important all aggressive prey be removed from the habitat. Following each molt the mantis may not be ready to eat for half a day to a few days.

It is possible to overfeed. Mantids will never eat so much they burst, but adult females will eat so much the abdomen will look ready to burst. Overly fat specimens should not be fed for a few days or more, to allow them to eliminate the last feeding as frass pellets. Nymphs generally cannot be overfed but if they are kept very fat and ventilation is inadequate, death is possible. Also, heavily fed empusid nymphs can be more likely to fall during a molt.

A regular feeding regime must be followed to ensure survival. Most other pet insects and arachnids can go weeks without food to no ill effect, but a week without food or water could wipe out an entire mantis collection. There are exceptions and certain prey items can be stocked or self-fed, but small nymphs should be attended to at least every other day.

Mantis eating a skipper. (CC) Aiko Vanhulsen

Mantis eating a cicada. (CC) Joi Ito

Too large of prey may cause the mantis to exhibit a threat display. It may still eat the prey, but usually just smacks it away. If this behavior is noticed, the size or type of prey should be changed. Too many prey will not cause such a strong reaction, but the mantis can be stressed if prey is constantly bumping into it.

Individual feeding behavior should be responded to appropriately. Hungry, aggressive mantids may run across the cage or feed on prey they normally would not bother with. This is only dangerous because it is easy to feed them something that might harm them, whereas less aggressive mantids would not attack something that poses the slightest threat. Picky mantids that refuse food may be induced to feed with flying or colorful prey. Mantids that go without food too long will not grow hungrier and feed—they will weaken and may lose the ability to catch prey. Flittering moths and green adult *Panchlora* are often fed on by animals that seem to have no appetite or would never eat something else as large. Once a picky mantis feeds a few times it may grow stronger and accept common prey or remain picky.

Mantids should not be fed just any old prey as if they were little garbage disposals. They can damage their jaws trying to eat overly hard prey or become ill from chemically protected prey. The common feeder pea weevils attract the attention of small nymphs, but the shell is too hard to chew for anything but Tarachodids. Mantids can become annoyed with certain food items and may even temporarily lose interest in food altogether if hard-shelled ants and beetles or chemically protected prey like milkweed bugs are used. Just two genera of cockroaches kept in the hobby, *Deropeltis* and *Diploptera,* produce quinones and should not be used as food. They are not usually sold as feeders. The poison does not seem to cause permanent harm but once a mantis eats one it usually refuses to touch any roach for a few days. Other, less potent, cockroach chemical defenses do not bother or deter mantids. Food caught outdoors is usually harmless as long as it is an orthopteran or lepidopteran, but if the prey item is sluggish it may have come in contact with a persistent pesticide. Mantids do not recognize pesticides so sluggish prey or feeders of unknown origin should be avoided. Hungry mantids may eat any animal small enough to capture and larger species have been photographed feeding on frogs, lizards, newts, mice, and small birds, but vertebrates are not a normal part of the diet (though Preston-Mafham 1990 reports when he was in Australia, he was told the prolonged screams heard some nights had been traced to frogs being slowly eaten alive by large mantids). In captivity, mantids are sometimes hand-fed such things as hamburger meat, banana, and hot dog. These unusual foods usually are not a problem for the adult mantis but they would likely adversely affect development if fed to nymphs as a large part of the diet. There are too many commonly available and readily cultured harmless feeder insects to justify feeding mantids food that may harm them.

Hand feeding ensures that mantids eat when you want them to and is necessary for old and deformed mantids which have little ability to catch their own food. Hand feeding may also be required for a mantis that is kept in a very large cage or is free-ranged on a houseplant. One method for hand feeding begins by tying a food animal to a thread and dangling it in front of the mantis until it strikes (kind of like fishing). The "fishing method" is often time consuming because the mantis sees you and may take a while to become accustomed to your presence. Another method is to hold the food item with forceps or impale it on a stick and offer it to the mantis. This "shoving food in the face method" is especially bad for timid or very old individuals because, not only does the mantis need to get used to your presence, but also, the forceps or thicker stick is seen as part of the food. Mantids will seldom strike if the prey appears too large. Another method is to hold onto the mantis just above the abdomen and shove food in its mouth as it tries to bite and pinch. A hungry mantis will become more interested as it tastes the food and then can be gently set down to continue the meal. The "making the mantis eat food instead of your hand method" is great in that it works on even the most docile and deformed mantis. Care must be taken not to squash the mantis and much time can be spent if it is more angry than hungry. Last of all, the mantis can be allowed to sit on a hand or arm while unanimated food is set in the other hand and gently offered to the mantis. Few mantids accept this last "hand feeding method."

For most species it is not possible to determine the exact diet each species is accustomed to in nature, but some genera are known to prey mainly on ants (e.g., *Gonatista*, *Liturgusa*, *Pyrgomantis*) (Gillon & Roy 1968; Ladau 2003; Schwarz 2003). Even if prey consumption were measured through detailed data collection, the results would range widely for the same species found in different regions or habitats and the natural prey might be impossible to culture. However, it is likely all species feed on a variety of prey in the wild so some variation in prey will help provide improved nutrition in captivity. Since exact prey nutrient levels (limited information is available on fat and protein content) and mantis species requirements have not been researched, it has only been possible to offer available prey items to determine what works best through trial and error. Fruit flies, crickets, and cockroaches have proven invaluable prey over decades, while various feeding results and considerations are recorded here. More recently, firebrats facilitated the breeding of *Eremiaphila* species that did not fare well in captivity previously.

Some hobbyists claim specific prey must be fed to certain types of mantids or they will die. The most common claims are empusids must only eat flies and ground mantids can never eat flies. There is a grain of truth to these statements because of the feeding style, but there is no evidence since both can be successfully reared on either. I had some trouble years ago feeding cockroaches and crickets to *Gongylus* long before it was possible for a hobbyist to find house flies for sale in any real capacity. (Maggots were rare at convenience stores that sold bait and the time to adult fly too long anyway.) Once I figured out how to adapt the cage and reduce the size of prey items, I was able to rear up *Gongylus* and later *Blepharopsis* without deaths. The prey has to be smaller and the cage perch (the lid) cannot be further from the bottom than the combined length of the animal and extended raptorial forelegs (so nymphs can simply grab prey running across the bottom). Empusids, especially *Gongylus*, generally will not walk around to hunt down food and would rather starve. Ground mantids are not built to catch flies or moths and flying prey can die without being eaten if a wing is not removed. Fruit flies lack in some amino acids which does not normally bother mantids, but causes severe growth issues for some creatures. Tarantula spiderlings develop "curly leg syndrome," while centipedes and whipspiders stop growing and cannot molt if fruit flies are used as the only food. Pupae of the house fly *Musca domestica* and other flies have become readily available in recent years and somehow this has led to claims the flower mantids must eat flies for success. Of course that was never true for

Mantis eating a dragonfly. (CC) Joi Ito

Do not feed a mantis with puffy wing-pads, as it should molt soon.

Ceratomantis sausurii with a house fly in each claw. © Kenneth Tinnesen

Mantids feed on whatever part of the prey is closest to their mouth.

countless breeders in the past and it is highly unlikely the hymenopodids all underwent unsolicited hyperevolution in the last few years. Except for *Mantoida* and a couple recent exotics lobster roaches have not yet been fed to, all mantids in this book have been successfully reared using *Nauphoeta cinerea* (after early instars have outgrown fruit flies). It is good to listen to feeding advice, but when someone claims pollen is necessary for proper development, flower mantids should only eat flies, or you must jump up and down three times and spin around before placing prey in the cage, weigh the chances of the information being true against the extra effort or money required.

House flies can be dusted with pollen or fed a liquid mix containing honey and pollen in an attempt to offset the amino acids flies are lacking. Bee pollen is mostly sugar and protein but contains tiny amounts of dozens of trace minerals. The trace minerals will be completely different depending on what pollens are contained and are only measurable in quantity. For this reason and since the mantis ingests very little relative to its diet, protein is the primary ingredient being employed. Protein powder has been used in its place and may also be mixed into fruit fly culture media. Some breeders swear by such methods but it is a recent fad. No experiments have been conducted to see if the pollen or protein is even making it to the mantis' gut and not being evacuated, let alone whether it affects development. The added protein may not help but it does not hurt either.

Prey can be dry dusted with calcium-vitamin powder to improve nutritive value. Commercially available powders that have been used are tailored to balance insectivorous reptile diets and contain mostly calcium, vitamin D, and a number of vitamins and minerals common in human multivitamins. Tiny amounts are used so a small jar can last a serious mantis breeder many years. Usage can be time consuming since prey must be dusted individually prior to each feeding, but is usually only employed in later instars when large prey are being fed. One informal study I reported (McMonigle, 2006), comparing a controlled and experimental group of *Hierodula membranacea* nymphs which were offered only *Nauphoeta cinerea* as prey, showed measurable results (the study included a concurrent experiment on a Mygalomorph with only slightly less impressive results). Each group contained twelve animals. All experimental animals came from the same ootheca. They were fed fruit flies until 3rd instar and kept in the same cage until the 2nd molt had been completed. After separation, the experimental group had all prey items dry dusted with the insectivore calcium/vitamin powder (Fluker's Repta-vitamin with beta carotene, 19.8% calcium) but otherwise caging and husbandry parameters were identical (marked cages located on the same shelf). None of the animals died in either group. Measurements were taken when specimens reached adulthood and consisted of length from the front of the head to the end of the abdomen using vernier calipers. Eight of the mantids in the experimental group were measurably longer than the control group by 30% at maturity. The other four matched the control group. This shows mantids can grow larger if a greater variety of nutrients are provided (or conversely that inadequate nutrition results in a reduced size). However, having slightly larger adults will not combat the main difficulties in mantis culture such as mating, time constraints, and proper ootheca gestation. It is possible calcium-vitamin powder abuse could cause abdominal blockage if glued on with water or honey in such a way that a massive quantity were ingested, but such an abuse would require extra effort and expense.

Mantis cannibalism at Irwin Prairie State Nature Preserve, Holland, Ohio. (CC) Benny Mazur

An entomologist friend for years would get small *Hierodula* and *Sphodromantis* nymphs from me once they had grown past fruit flies. He fed strictly with crickets from the pet store. When I'd see his adults months later they were 2/3rds the size of mine from the same ootheca though he kept them in much larger cages. Crickets are not a bad food, but they are certainly not as nutritious as common feeder roaches. Prete recorded development times for *Sphodromantis lineola* and noted that around 20% of his nymphs would take six months instead of three. He theorized it was a natural developmental bifurcation of the population (Prete et al. 1999). However, I have never seen anything like that in *Sphodromantis* or any other species, and he only mentions using crickets after fruit flies, so his stunted animals were likely the result of poor nutrition. (Many insects, especially rhinoceros beetles, with more nuanced food requirements develop close together when fed quality substrate but can come out years apart if provided inadequate nutrition.) Nymphs from the same ootheca always develop at different rates but rarely vary by more than 20-30% of the development time, not 100% under identical conditions.

Providing praying mantids with a variety of feeder insects can make them healthier and larger and allows females to produce more and larger oothecae. Keeping cultures of a few different feeders on hand is very useful since prey can be switched up in cases where mantids refuse the prey being offered or to reduce the chances of running out of food due to the normal cyclical nature of prey cultures. Most cultures can be maintained for months or years on food that costs pennies, while a few dozen crickets or a fruit fly culture from the store can cost five or ten dollars every week. The cost adds up even if very few mantids are being kept. Some prey cultures require no time investment, while others are labor intensive. Many of the prey animals listed are readily available through local pet shops and stores that sell bait, while sources for various starter cultures can be found online.

The following list includes pros, cons, and all the information necessary to rear a number of insects commonly available as food for mantids. An entire chapter could be written on variations for each, but this is a book about the praying mantis.

1. House crickets *Acheta domesticus*. (*Acheta* is masculine, even though it ends in –*a*, so the specific name must end in –*us*. Most breeders and textbooks get this wrong). The house cricket is the most common feeder. The stock pictured are a hypermelanistic (black) form of the normal feeder *A. domesticus*, cultured to hatch out at lower temperatures. They are not field crickets (*Gryllus* spp.), which have short rear wings, black nymphs, and are much more difficult to rear.

Pros: These are active, time-tested prey that quickly draw the attention of hungry mantids. Feeder crickets can be purchased at nearly any pet shop and through the mail. They are commonly available in numerous sizes to accommodate the size of the mantids being fed.

Cons: Crickets hide and cannot fly or climb glass. Uneaten crickets are extremely dangerous, especially to molting mantids and oothecae. If not eaten right away, they can die and smell. (As luck would have it they never seem to die if the mantis is going to molt or make a fresh ootheca.) Except for bulk purchases, crickets can be very expensive to purchase over the long run. Rearing large quantities is space and time intensive. Crickets are excellent jumpers and often escape into the home where they can chew on books, food, and other personal items. The sound of a loud male chirping

House crickets.

Wingless *Drosophila melanogaster*. © Henry Kohler

away every night interrupts the sleep of many hobbyists' spouses. Small crickets, especially pinheads, are very difficult to handle and those that do not jump away are often squashed by the act of picking them up.

Rearing Instructions: Adult female house crickets lay a few hundred eggs in any potting soil or dirt that is an inch or more deep. Removable trays can be relocated to new cages if varying sizes need to be isolated for different tasks. The dirt should be kept moist until tiny pinhead crickets hatch out after a few weeks. The nymphs and adults do well on many foods such as leaves, leftovers, old fruits and vegetables, and dry dog or cat food. Water in a sponge or gravel filled dish should also be provided to the growing crickets. It is necessary to use a screen or otherwise ventilated cage or the crickets will die within a few hours to a few days. Cage decorations consist of paper egg carton for them to molt and hang out on. Additionally, at least a 60-watt incandescent light bulb located directly atop the cage is a necessity if the ambient temperature is below 75° F. Without this source of heat it is often impossible to rear a single cricket because the eggs will not hatch.

Feeding: Crickets are usually kept in large tubs or terraria, so trying to catch them out one at a time is unrealistic. Shake the egg carton over a tall deli-cup and catch the desired size from this container. Pinheads can be poured in through a funnel.

2. Fruit flies. *Drosophila melanogaster* and *D. hydei* are the primary prey used to feed early instar mantids. *Drosophila hydei* are two to three times the mass of *D. melanogaster* and a little touchier.

Pros: Fruit flies are available in wingless varieties and in two sizes. Apterous and flightless (some varieties have wings but they do not work) varieties are easy to feed. Fruit flies can be purchased through mail order nearly anywhere in the U.S. Fruit flies do not cause damage to young mantids. Wild fruit flies are easily procured at no cost. A small shelf devoted to fruit flies is all that is needed to feed a hundred or more hatchling mantids in the first few instars.

Cons: Fruit fly cultures can be quite expensive and orders usually arrive long after unexpected hatchling mantids starve. Flies can die before being eaten. Even tiny mantids need to be fed a lot of fruit flies. After a few molts most mantids are too big to bother with even the largest fruit flies. Fruit flies are smashed easily by a person's hands. Cultures go through boom and bust cycles. A number of cultures must be maintained and new cultures started every week or less if the hobbyist wishes to have them on hand at all times. Keeping an adequate supply of fruit flies at all times can require some effort. Adult fruit flies live only a few days and often die before being eaten. Too many flies or big chunks of the substrate often fall out when feeding. Pre-made media is expensive. Wild fruit flies loose in the house can be very annoying.

Rearing Instructions: Fruit flies are kept in a sealed container with a breathable lid and a half an inch of media in the bottom. The lid keeps in *Drosophila* and keeps out phorid flies. Small 1.5" diameter vials with a sponge stopper or a coffee filter rubber-banded over the top used to be the accepted standard. The tiny cultures were very difficult to maintain and moderate production required dozens of containers. Today we use 32oz. tall deli-cups with the coffee filter built into the lid. These large containers are easier overall and two or three in series provides enough flies for hundreds of mantids. The media can be made of numerous items including oatmeal, banana, apple, etc., but the easiest and cheapest consists of instant potato flakes and water (the exact amount depends on ambient humidity, but about 1/2 cup flakes and slightly more than 1/2 cup water for one 32oz.). The potato should stiffen up within a minute of being mixed or more flakes may be needed. If media is too watery it will drown the adult flies before they lay eggs or pour out when the container is tilted for feeding. A pinch of yeast is added to stimulate egg laying, as well as increase success and growth. Conversely, a dash of liquor, vinegar, or apple slice will work. It is important to keep cultures in series because they usually exhaust after a few weeks even if all the food is not used up. With the large containers, setting up a new one every two to three weeks provides continuous flies, while with little vials, half a dozen had to be started each week. Remove about three dozen of the first wave of flies to start a new culture. If a culture becomes overrun with mites, it has been kept around too long. Larvae will eat slime bacteria and mold that starts to grow on the surface about a week after the culture is started. When the first pupa is seen on the side, throw in a crumpled paper towel so pupae can form on it or most will drown and die. If added too early, the paper can reduce or prevent egg laying. (Excelsior, which is long, very thin strips of aspen wood shavings, or corrugated cardboard tubes can be placed earlier since they do not affect egg laying as much). Mold inhibitor is not necessary and may be impossible to acquire these days with increased regulation. Mold only overgrows cultures that were started with too few flies, not provided an egg-laying stimulus, or made too dry. The temperature should be from 75-85° F. Warm temperatures shorten the cycle time by a few days, while conditions that are too cold can stop production. Wild fruit flies are easily attracted using banana, apple, or other fruits. A large number of wild fruit flies can be attracted or reared in a short time. More rotten fruit smell attracts more flies. Normally, containers are placed in the garage or a shaded spot outside, but a fruit bowl by the kitchen window works, too. A container with a number of small holes will trap flies or the lid can be strategically placed nearby.

Feeding: Use of the wingless or flightless variety is simple: the bottle is held partly upside down and tapped lightly with the palm until the appropriate amount of flies is fed. Wild fruit flies can be fed in the same way as flightless species but must be placed in the freezer for five minutes. The flies' body functions shut down from the cold, but pay attention—they warm up quickly! A small apple slice may need to be placed in the mantis cage to keep the flies alive for a week or two. Otherwise they may dry out within hours.

3. Cockroaches. Feeder roaches are employed most often by serious hobbyists or mantis breeders, rather than someone who keeps one or two pet mantids. There are twice as many cockroach species as mantids. More than a hundred species are kept in captivity, many for decades, with at least one stock dating back to the early 1940s. They are highly variable: some climb glass, some cannot, some burrow, some do not, and some are gigantic (or tiny) at first instar. However, they are mostly easy to care for, useful, and can be reared on cheap dry dog food. Some are nearly impossible to keep, but many breed like roaches.

There are a few main types and sizes of roaches available for feeding mantids. 1. Large, non-climbing species like *Blaberus*, *Blaptica*, and *Eublaberus* are unable to climb glass, which makes them easy to handle. Even first instar are the size of a house fly and the mantis may not want to chase them to the bottom of the cage. 2. Lobster roaches *Nauphoeta cinerea* and firefly mimics *Schultesia lampyridiformis* are more appropriately sized, are born small enough to feed some L1 mantids, do not burrow, and climb glass. One decent *Nauphoeta* colony in a five-gallon bucket cage can provide enough meat for hundreds and hundreds of mantids. 3. Green banana roach *Panchlora nivea* adults fly and climb and are highly attractive to some mantids, but the nymphs burrow and do not climb smooth surfaces. 4. *Dorylaea orini* and *Eurycotis decipiens* are high profile climbers that can be very productive and make great feeders, but culture starts are very expensive. 5. *Blatta lateralis* (Turkestan roach) breeds readily and does not burrow but cannot climb.

Dorylaea orini.

Pros: Keeping roaches is neither time intensive nor expensive. Cockroaches live inside mantis cages without food or water until the mantis is ready to eat them. Roaches do not bother molting mantids (even the rare species with predatory habits will not climb to eat a mantis). Only one or two roaches of the appropriate size are needed for the mantis to make it to the next molt or to lay the next ootheca. One established culture (lobsters especially) in a five-gallon bucket or plastic tote can provide endless mantis food with minimal care.

Cons: Even first instar nymphs of most species are too big to feed small mantids. Many species are normally offered only as starter cultures and some are expensive. Exotics, primarily livebearers, are normally kept because they cannot infest homes—our native woods species are not a viable option because they do not produce in quantity. There are five common pest species that might work well, but are almost never used because they can escape during feeding. (The small *Blattella germanica*, the most prolific of pests, would probably be a great feeder, but I'd never bring one to my home or lab.) Roaches hide well in decorated tanks and some roaches even hide well in bare bottom cages. Mantis oothecae left with

Lobster roaches.

certain, large, hungry roaches may get chewed on. Large cages can be necessary to rear large species.

Rearing Instructions: Roaches are often kept in five gallon buckets, plastic totes, or fifty-gallon trash cans. Like mantids, it is important to have some screen ventilation or they die. Egg cartons or pieces of rough wood or bark are placed in the cage for climbing and hiding. A water dish containing gravel or sponge to prevent drowning should be provided or an area of damp substrate. Nearly any human or pet food can be used to feed roaches. Fruit such as apple cores or melon rinds should be offered from time to time to the live bearing genera (i.e., *Panchlora*, *Blaberus*, *Eublaberus*, Schultesia, and *Nauphoeta*) to improve production. Egg laying genera should be provided cork bark that is misted from time to time to keep the oothecae from drying out (i.e., *Eurycotis* and *Dorylaea*). Cages that contain glass-climbing species should have the upper two inches coated with petroleum jelly to help prevent escapes when the lid is removed.

4. Yellow Mealworms *Tenebrio molitor*. Mealworms are an easily accessible and non-dangerous food for mantids.

Pros: Mealworms are among the most accessible foods and can often be found at grocery, convenience, and fishing/tackle/bait stores, as well as pet shops. Hatchling larvae are half the mass of a fruit fly and can be fed to the smallest baby mantis. Mealworms pose no threat to the molting mantis or its ootheca. They are inexpensive and there is almost zero care involved. Relatively huge cultures can be reared in a small space.

Mealworms

Cons: It can take months or years to establish a large colony for feeding. Tiny larvae are a possible but not realistic food for early instar mantids. Although many mantids eventually come to the bottom of the cage to eat the larvae, some mealworms seldom move and never attract attention. A large number of full-grown larvae are needed to provide the nutrients for a big female mantis to produce a single ootheca.

Rearing Instructions: Any glass or plastic container with a screen lid will suffice as a cage. Fill the container a few inches from the top with plain, uncooked oatmeal or chicken mash. Throw in beetles or larvae. Dry dog or cat food can be tossed on top or used once there is a layer of frass. Do not provide moisture, do not

Blue bottle fly.

Blue bottle fly maggots and pupae.

mist, and do not add potato slices, or a grain mite infestation can occur and require the culture to be thrown out. A solid lid can be used on the rearing container but moisture build up could cause the substrate to mold. Pupae are usually moved to a separate container for best production, as this stage is often cannibalized.

5. Meal Moths *Plodia interpunctella* and others. There are some similar looking meal moths, and the large *Pyralis farinalis* which is similar in growth and care. Caterpillars make silk tubes and look similar to young waxworms.

Pros: Rearing takes minimal space and food is inexpensive. Meal moth cultures are unlikely to crash with little attention. Adults and caterpillars can be used as food. Mantids enjoy catching the flittering adults. Adult males are small enough to make good food for many hatchling mantis species even if they are much larger than the L1 mantis.

Cons: Cultures can take many months to become established and still are often sporadic or seasonal. If pantry foods are not sealed, cultures can become much larger than desired. (*Plodia* especially are aggressive household pests of improperly stored plant products, mostly kernels and seeds.) The moths are difficult to handle. Full-grown caterpillars wander before making cocoons, so rearing containers must be sealed.

Rearing Instructions: Adults or caterpillars should be placed in a sealed and vented plastic or glass container. Wild birdseed can fill the container nearly to the top. Additional moisture is not helpful and the only care involves replacing food a few times a year.

Feeding: Cocoons are usually made in the upper edges of the container and can be removed and placed in mantis cages before the adult moths emerge, but this can take weeks. Moths are easy to smash when handled, though the caterpillars are very sturdy.

6. Blue bottle flies, house flies, stable flies, etc. (Family Muscidae). The house fly *Musca domestica* is available in flightless cultures but is primarily used because it flies.

Pros: House flies are an excellent food for feeding larger mantids that prefer not to hunt and can reduce cannibalism because they elicit a different hunt response from crawling prey. Refrigerating house flies doesn't work, but does work well with maggots of the flesh flies (*Calliphora*, *Lucilia*). These maggots are sold as fishing bait.

Cons: Loose flies are difficult to catch, though blue bottles usually end up at a window where they are easy to recapture. Pupae can take a few weeks to hatch and purchased pupae do not have the best hatch rates. Uneaten house flies must be fed in mantis cages or they to fall over dead in a day or two. House flies are very active and can dislodge even a surprisingly large, molting mantis, so excess prey may need to be removed if uneaten. Rearing is a very smelly proposition and is more labor intensive than for most feeders.

Rearing Instructions: House fly media consists of dry dog food saturated in water or milk (with milk being a necessity for blue bottle media). Adult flies will lay eggs on the media and in a few weeks there will be plenty of large maggots. There should be a small container of moist dirt in the cage for full-grown larvae to climb in and pupate or they can be moved out as they mature.

Feeding: Like meal moths, pupae are easily removed and placed in the mantis cage before the adult flies emerge. A large number of pupae can be placed in a pop bottle to hatch and then fed off by placing the mouth of the bottle inside the feeding hole.

7. Humpbacked flies. Family Phoridae, commonly known as phorid or "scuttle" flies due to their unusual (for a fly) habit of running across surfaces.

Pros: Phorid flies grow much more quickly and are far more hardy than fruit flies. Humpbacked flies are more variable in size than fruit flies. Phoridae walk as much as fly and are more easily smashed if they escape.

Cons: Phorid flies are not presently available in a flightless form. Larvae feed on wet decaying matter that includes the bodies of freshly dead arthropods. Phorids infest cultures of crickets and roaches unless dead are removed immediately. Dead insect bodies are crawling with maggots in a few days. (The incredible speed of reproduction leads some people to falsely assume phorids kill invertebrates.)

Rearing Instructions: Phorid flies can be kept the same as house flies or fruit flies, but it is quite difficult to drown or desiccate pupae. Larvae wander and tend to pupate in protected spots or up by the lid.

Feeding: As with wild fruit flies, culture containers are placed in the freezer for five minutes to temporarily immobilize adult flies for feeding to mantids. If phorid flies have already become a pest in other food cultures, dead insects can be thrown into a screen cage with young or hatchling mantids. Loose flies will be attracted into the cage by the dead bugs—this will not get rid of the phorids but it is a great way to keep small mantids well-stocked with food.

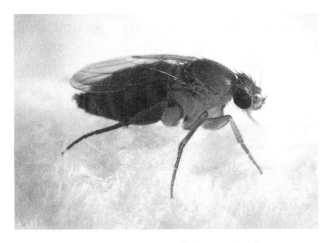

Phorid fly up close. © Henry Kohler

8. "Giant springtails" (Order Collembola). There are a number of springtails commonly cultured for use with dart frogs, but only the

"Giant" springtail culture.

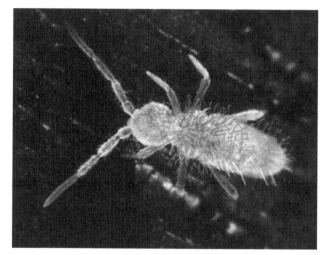

"Giant" springtail up close. © Henry Kohler

common species sold under this trade name is really useful for mantids because it stays on the surface and the maximum size is relatively large (as big as a very small *Drosophila*).

Pros: Springtails are tiny and can be a great starter food for even the most timid of hatchling mantids. Easy and nearly free to culture.

Cons: They cannot climb clean glass. It takes a few months or more to establish a large culture. Since they are very tiny, tons of springtails are necessary to feed all but the smallest mantis. Springtails are too small to feed nymphs of appreciable size. They are far too delicate to handle directly and dry out more rapidly than fruit flies. Likewise, a piece of apple in the cage helps keep them alive.

Rearing Instructions: Springtails feed on a variety of different foods but are normally raised on fish flakes, dog food, and fruit. Very little food is required. Springtails are kept on a shallow layer of potting soil, compost, or charcoal and do best if kept in medium to high moisture. They require little ventilation so a few pinholes or the normal cracks around a shoebox lid provide plenty of air. Remove any small brown mites that appear in the culture container or the production will be minimal or negative (the mites eat them slowly).

Feeding: Paper egg carton or chunks of wood are kept in the cage, which are removed and shaken over the mantis's cage for feeding (springtails congregate on flat surfaces).

9. Waxworm moths *Galleria mellonella* and *Achroia grisella*. *Pros:* Commonly available as bait, adults are great flying prey for picky eaters like *Gongylus* and *Idolomantis*. Easy to culture in mass. Can be refrigerated and used as needed.

Cons: Timing is essential because they are available only as larvae, so it is necessary to wait a number of weeks for mature larvae to pupate and then moths to emerge. Waxworms are known for high fat content and minimal nutrients. If the larval diet consists of more than a trace of honey it can be very expensive to feed the caterpillars.

Rearing Instructions: Like house flies, different containers are needed for adults and pupae and another for the media, though the separate containers can be kept together inside a large plastic tub. The moths live a week or two and will lay eggs on strips of wax paper because it feels like comb wax. The paper is placed on top of the culture media for hatching. Larvae feed on honeycomb outdoors but

Waxworms.

Waxworm moth.

Waxworms and moths.

Silkworms.

can be offered honey mixed with dry dog food, cereal, or glycerin in culture. Larvae start out tiny and may escape and chew up any nearby object made out of wood or paper if not properly contained. They cannot climb smooth, clean surfaces or culture containers can be fully screened (metal microscreen only, as cloth or paper screen is no barrier). Mold and grain mites may decimate cultures with few larvae or those kept too cool (around 80° F is suggested but not required). A piece of paper towel or egg carton can be placed in the cage to give mature larvae a spot to form loose, barely visible cocoons. Note that waxworms from bait shops are normally sterilized and eggs procured from them will not hatch.

10. Silkworm moths *Bombyx mori*. *Pros:* Some mantids prefer caterpillars to other foods and will take larger caterpillars than other prey.

Eggs can be refrigerated until needed. Caterpillars are easy to handle. Larvae and eggs are readily available by mail and are not very expensive. Food for caterpillars is free if you have a mulberry tree nearby.

Cons: More time will be spent caring for the silkworms than the mantids they are being fed to. Caterpillars produce a lot of waste. Hatchling caterpillars dry out easily and do not make good food for tiny mantids. The life cycle takes a few months, so much effort is required to match predator and prey size. They often refuse to move and may never attract attention. Pre-made and artificial diet is expensive in small quantities.

Rearing Instructions: Caterpillars feed on mulberry leaves or artificial diet. Artificial diet comes in powder form (with directions for preparation) or pre-made blocks. Cages for silkworms must be open air or screen—a shallow tray works well since growing caterpillars rarely wander off. Fecal pellets should be removed any time they build up enough to retain moisture and become moldy. The moths are rarely used as food since they are too large for most nymphs, cannot fly, and often do not live a week. Eggs are laid on paper or petri dishes and may not hatch unless refrigerated for a few months, depending on the day length the parents were reared under.

11. Firebrats *Thermobia domestica*. *Pros:* These are primarily indicated for feeding to *Eremiaphila* and other mantids that do not thrive on fruit flies. Once established, cultures are very productive. They are harmless to mantids.

Cons: They cannot climb glass and mantids often ignore them. Starts are expensive and can be difficult to acquire. Culture containers must be adapted with special heating due to the high temperatures required for reproduction and there are plenty of alternatives.

Firebrats.

Rearing Instructions: Firebrats are kept on paper egg cartons and fed dog or fish food. Cotton balls are provided to offer egg-laying spots away from the hungry maws of the adults and immatures. They are very easy to keep alive and grow to maturity, but getting good production is not so simple. High temperatures (>90° F, optimum 102° F) required.

12. Isopods. Non-volvating species. Volvating species (pillbugs) tend to be difficult for the mantis to grasp or bite.

Pros: These again can be useful primarily for some ground-dwelling species. Isopods can

White micropod *Trichorhina tomentosa*.

be collected in most areas and from under logs even when there is a foot of snow on top. They are easy to culture and have no unique requirements. "Micropod" species like *Trichorhina tomentosa* and the smaller Costa Rican micropod start out smaller than a fruit fly eyeball. This is one of the few terrestrial arthropods with a calcified exoskeleton, but they are relatively soft.

Cons: They cannot climb glass and mantids often ignore them. They desiccate rapidly or hide under substrate. They have been used to supplement the diet but there is no record of use of these as a singular or primary food. Even large *Porcellio* are small compared to an adult cricket. Reproduction is rather slow compared to *Drosophila*.

Rearing Instructions: An inch or so of damp potting soil is placed in a plastic shoebox or bucket with a lid. A few holes can be added to the lid for ventilation and some brown, decayed hardwood leaves and pieces of wood or bark placed on the substrate. Dog food, fish flakes, or fruit can be added to supplement the diet, but without the leaves and wood females seldom produce young.

Large mantids have been known to take hummingbirds. (CC) JeffreyW

USA MANTIDS

There are 28 or so species found in the continental United States. In many areas the most commonly encountered mantids are not native, but were introduced in the late 1800s to early 1900s (marked with *). Nine of the listed species have rarely, if ever, been kept by hobbyists (marked with †) and some of these may be unverified synonyms for common species. Since the large adventives prefer disturbed areas, they are the species most commonly encountered by people. Many of the native species are more difficult to find, since they are smaller and often blend in well with native flora or are uncommon. Hawaii, which has no native mantids, hosts four members of the Family Mantidae: *Brunneria borealis, Hierodula patellifera, Tenodera australasiae,* and *Tenodera angustipennis*, according to specimen holdings at the University of Hawaii at Manoa Insect Museum. (The middle two would bring the U.S. list up to 30, though Howarth & Mull 1992 state there are six total species adventive to Hawaii.) Continental ranges not directly cited are according to Helfer (1963), with some adjustments made for recent collection records.

FAMILY ACANTHOPIDAE
Tithrone clauseni†
Tithrone corseuili†

FAMILY MANTOIDIDAE
Mantoida maya†

FAMILY THESPIDAE
Bactromantis mexicana
Bactromantis virga†
Oligonicella bolliana†
Oligonicella scudderi
Thesprotia graminis

FAMILY TARACHODIDAE
*Iris oratoria**

FAMILY LITURGUSIDAE
Gonatista grisea

FAMILY MANTIDAE
Brunneria borealis
Litaneutria borealis†
Litaneutria longipennis†
Litaneutria obscura
Litaneutria minor
*Mantis religiosa**
Phyllovates chlorophaea
Pseudovates arizonae
Stagmomantis californica
Stagmomantis carolina
Stagmomantis floridensis
Stagmomantis gracilipes†
Stagmomantis limbata
Stagmomantis montana†

Tenodera angustipennis*
Tenodera sinensis*
Yersiniops solitarium
Yersiniops sophronicum

The greatest aspect of U.S. mantids (for U.S. hobbyists) is that they can be collected from the wild or acquired from other local hobbyists. Unfortunately all our largest species are much more difficult than many tropicals for reasons specific to each species. Some of the small species like *Thesprotia graminis* and *Oligonicella scudderi* are seldom kept but are easy to rear through generations.

Adults are collected in mid- to late summer. The tiny nymphs, found early in the season, can be much more of a challenge to locate. Large species are easiest to find on sunny days in gardens or meadows located in full sun. The vegetation can be swept with a large net in order to catch hidden grass mantids. Ground mantids can be sent running into a waiting net on the ground. Flying species often come to lights after dusk and can be collected on store windows or walls, all-night gas stations, tennis courts, parking lots, or under sills and ledges. Well-lit strip malls can be exceptionally good places to find mantids. Males of small natives like *Oligonicella*, *Stagmomantis*, and *Thesprotia* may be found in numbers at lights, but the flightless females have to be hunted in the field. Immature females will need to be mated once they are old enough, but wild adult females have already mated.

Native oothecae can be easier to collect than mantids. Sometimes it is only easy to collect oothecae in late fall after the leaves are gone and the foamy globs are easy to see from a distance. It is not rare to see none in the summer but find dozens or hundreds in the fall. Many of our native species produce smaller oothecae pasted low on branches and rocks or within a foot of the ground on brick buildings. Oothecae can be hatched and young raised to adulthood at any time of the year. If natives are to be released for pest control it is important to synchronize emergence with spring or they will face out of season death.

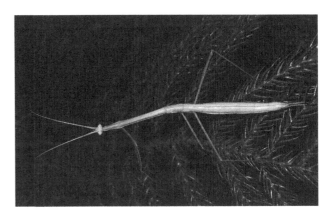

Brunneria borealis green coloration.

Brunneria borealis Brunner's Mantis
Although the Brunner's is not the most massive U.S. mantis, it is the longest, growing about a half inch longer than our biggest adventive *Tenodera*. The abdomen is variably striped in yellow, red, and green. Wild specimens have a base color of grass green while captive reared specimens usually have a base color of pink (either are possible). The tiny useless wings of the adult are a shade of pink or purple accordingly. The unique feature that sets *B. borealis* apart from other U.S. mantids is the shape of this single-gender species' antennae. The female's antennae are thick and lightly bipectinate (male antennae of other *Brunneria* species are similar but longer). Sometimes this species is referred to as a stick mantis, but *B. borealis* lives in tall grasses and looks nothing like a stick. I have seen a few in the wild and it is unbelievable how difficult it is to see one among tall grasses even if staring right at it. Their natural reproductive method sets Brunner's apart from every other mantis

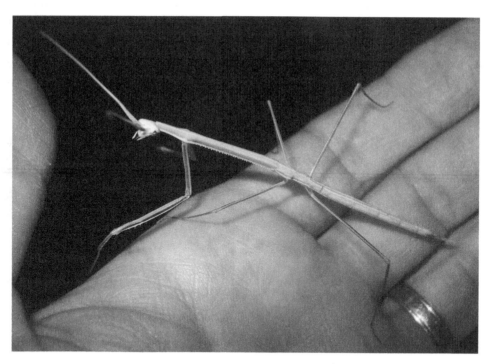

Captive-reared *Brunneria borealis* pink coloration.

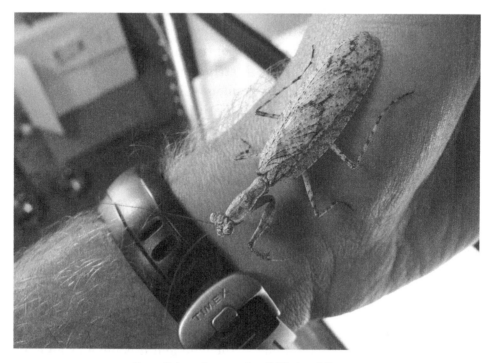

Gonatista grisea male. © Richard Trone

in the world. Because there are no males, females reproduce through parthenogenesis. Females commonly produce ten or more oothecae that contain twenty to forty eggs each. Oothecae are half an inch long by a quarter inch wide. Of course *B. borealis* is the unique feminist and has decided not to follow the usual ootheca structure with a large hatching zone. She creates an ootheca where every hatching nymph must come out individually from a single aperture at the end of the "spine." The normal hatch time at room temperature is four to six months, not the usual four to six weeks. Two to three dozen nymphs emerge on different days, up to two months apart from a single ootheca. Keeping the eggs cool (40-50° F) for two to three months delays hatching time by a few months and causes nymphs to all hatch out within a few days. (Refrigerating an ootheca after it has been gestating at room temperature for a few months may prove deadly.) The greatest problem in rearing these is getting the newly hatched nymphs to eat. Though the first instar nymphs are bigger than most hatchlings, it can be difficult to get them to start on small fruit flies. They can starve unless "giant" springtails, or other tiny foods are made available. Once the mantids get past that first molt, they will take the smallest fruit fly species. Nymphs and adults prefer food animals that are a maximum of three times the size of the mantis' head. *Brunneria borealis* is found throughout the southern U.S., from Texas to North Carolina and Florida, and is adventive in Hawaii. The pictured animals are from Georgia and North Carolina.

Gonatista grisea Grizzled Bark Mantis
The grizzled mantis is a member of the family Liturgusidae, the lichen or crab mantids, which look very similar across related genera. Nymphs and adults are heavily mottled to

Gonatista grisea 1st instar nymphs.

match lichen covered bark. They can move very quickly, darting back and forth and side to side around tree trunks in a manner that resembles the movement of certain crabs. Males are fully winged, while female's wings only stretch two-thirds of the way down the abdomen. Females have a lobed abdomen and look similar to the Puerto Rican species but are darker and have beautiful tiny shimmers of metallic highlights. Adults range from an inch and a third to an inch and two-thirds in length. The nymphs and adults are broad and flattened. When they press themselves against the bark of a tree, they literally disappear. This mantis holds its front arms out to the side and also runs sideways. The flattened, triangular oothecae contain thirty to forty eggs. This is not a beginner's species because the nymphs are difficult to take care of in the early instars. Hatchlings should never be allowed to reach a rough, horizontal surface like a paper vented lid because they hang themselves by the rear legs where they dangle uselessly, starve, and dehydrate, since

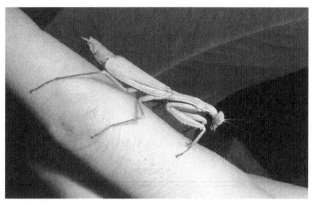

Iris oratoria female, stock from Arizona.

they cannot right themselves with the middle or front legs and will not let go. Grizzled mantids can be found in the southeastern U.S. as far north as North Carolina, but captive specimens have invariably come from Florida.

Iris oratoria Mediterranean Mantis
At first glance *I. oratoria* closely resembles our native *Stagmomantis* species in size and shape, but it is an adventive, introduced to the U.S. in the 1930s. Like *Stagmomantis* spp., females have shortened tegmina while the male's wings stretch past the end of the abdomen. Each hind wing of the adult has a large black eyespot. *Iris oratoria* nymphs are differentiated from *Stagmomantis* nymphs by the presence of six spines on the outer front tibia instead of five, and two tiny knobs on the front of the face between the eyes. There is also a small marking on the posterior ventral surface of the abdomen useful for identifying small nymphs. Adults range in size from one and a quarter to one and three-quarters inches. At first, hatchlings feed timidly and can require tiny food such as hatchling mealworms or springtails. After the first molt, the nymphs become superb eaters and devour baby roaches and small flies. The first three to four instars require substrate in the rearing container to conserve moisture or they experience mass die-off.

Adults and nymphs prefer foods that are less than half their mass. Oothecae are usually one inch long by a quarter inch wide and are triangular in cross-section. Helfer (1963) indicates that the Mediterranean mantis was introduced about 1933 and was established only in southern California north to San Jose where it was found around cotton fields. As of 2012 I have seen *Iris oratoria* specimens from much of the Southwest, specifically Arizona, Nevada, New Mexico, and Texas.

Litaneutria minor male. © Richard Trone

Litaneutria minor Minor Ground Mantis
Females can display a variety of colors and patterns at adulthood. This species reaches only about 1.3" at maturity and is normally found on the ground rather than climbing vegetation. They cannot climb glass and usually form the ootheca against the bottom of the cage but will form it on the screen lid if given something to climb. Each can contain 50 or more eggs and take months to hatch. Mature males usually have long wings but specimens with short wings are not unheard of. The specimens and oothecae depicted are from Arizona. They are known from dry scrub areas and the natural range is from British Columbia to North Dakota, south to Mexico.

Iris oratoria threat display. © Wolfpuppy

Litaneutria obscura female and ootheca. Specimen courtesy Richard Trone.

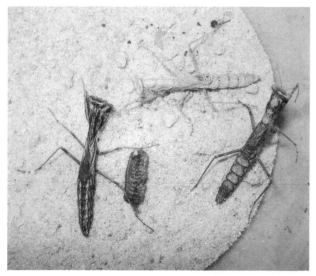

Litaneutria minor, just a few of the possible color patterns. Arizona specimens.

Litaneutria obscura and *Litaneutria minor* ootheca comparison (both average).

Litaneutria obscura Obscure Ground Mantis
The adult female of this mantis is just a hair smaller than *L. minor* but the oothecae average half the size and probably never contain more than a dozen or two eggs. Oothecae are nearly black (as opposed to the orange *L. minor* oothecae) and more narrow in construction. This species is found in California.

Mantis religiosa European Mantis
This is the second most commonly encountered North American mantis and also the state insect of Connecticut. The European mantis is a bit smaller than our *Tenodera* species with females averaging two and a half inches. Once I was sent a wild-caught adult female from Arizona that threw me on identification at first because it was barely an inch and a half long. This species is easily differentiated from our other species by the yellow "eye spot" with a black circle around it on the inner surface of each foreleg, close to the point of attachment to the thorax. Some individuals, however, have a dark blue area instead of the eyespot. The face lacks vertical stripes and the hind wings are transparent and colorless. The ootheca can be

Mantis religiosa, two oothecae.

surprisingly large, and looks different from a Chinese mantis ootheca because it is more tapered. The ootheca is not flattened and does not have vertical grooves like the one of the narrow-winged mantis. It is larger and more foamy than *Stagmomantis* oothecae. Berg et al. (2011) state it is preferentially formed under stones and rocks, though every Ohio oothecae I have seen was formed on thin herbaceus stems or grasses. This differs from *Stagmomantis* oothecae that are formed on thick twigs or flat surfaces in the wild. There are eight *M. religiosa* subspecies found across the Old World and ranging into the tropics (Berg et al. 2011) so not all oothecae require a cold period.

KEEPING THE PRAYING MANTIS 129

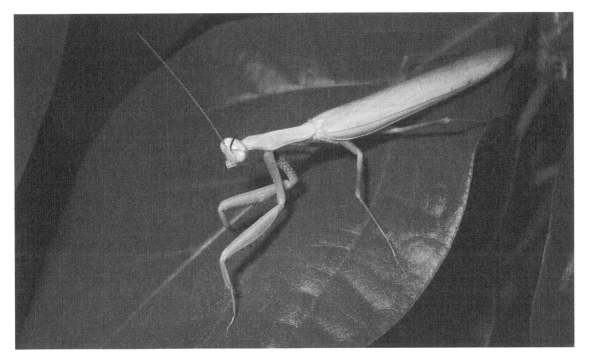

Mantis religiosa green.

Mantis religiosa brown.

However, chances are you will never see a tropical subspecies, so oothecae of the European mantis must be exposed to cold for a few months or they will not hatch. Two or three months in the refrigerator or outside during the winter in temperate climates is a sufficient length of time.

Mantoida maya Little Yucatan Mantis
This miniature mantis grows to, at most, two-thirds of an inch. The nymph resembles an ant all the way through subadult. The subadult pictured was found running across a biking path in Jonathan Dickenson State Park, while even adult specimens are seen scurrying across the forest floor. This species has a tiny U.S. range and has not yet been introduced to the hobby. This might be among our only species capable of overlapping generations in nature (though *S. carolina* is in Ft. Lauderdale, FL, according to Hurd (1999)) but there is no information for diapause of nymphs or oothecae as the details of its life cycle are undocumented. It is found in southern Florida and the Yucatan.

Oliginicella scudderi. © Yen Saw

Oligonicella scudderi adult female

Oligonicella scudderi Scudder's Mantis
The Scudder's mantis is among the smaller U.S. mantids at one to one and a third inches long. The body is delicately striped lengthwise in brown and tan or can be solid brown. Adult females have undeveloped wing pads while males' wings stretch nearly to the end of the abdomen. *Oligonicella* are easy to distinguish from other U.S. mantids by the presence of two small, but noticeable, bumps at the back of the head behind the eyes. Nymphs and adults feed well, but only accept food animals that are a quarter the mass of the body or less. Hatchlings are incredibly tiny and require springtails or severely stunted *D. melanogaster*. Small nymphs do not eat each other and can be kept together safely in small cages until at least the fourth molt, if not to adulthood. Females produce ten or more oothecae and can form as many as three in a single night. Oothecae contain just 10-20 eggs and are a quarter inch long by an eighth inch wide. Gestation at room temperature is about thirty days. *Oligonicella scudderi* can be found from Nebraska to Florida. *Bactromantis* (*Oligonicella*) *mexicana* is slightly longer and has a noticeably more elongate rear section of the pronotum. It is found from Arizona to Central America.

Phyllovates chlorophaea Texas Unicorn Mantis
This handsome species is among our largest and bulkiest mantids—females can exceed two

Mantoida maya subadult. © Tony DiTerlizzi

Texas unicorn mantis nymphs, Estero Llano stock.

and three-quarters inches. Despite visible differences, our two unicorn species are similar even down to the angle of the markings on the tegmina. It is amazing they are placed in separate genera. *Phyllovates chlorophaea* does not have notable lobes on the legs and abdomen. Immatures that hatch at the same time can be reared together with few to no losses. Nymphs are sturdy and can take a lot of abuse, but can suffer from mismolts in the final instar with improper molting surfaces or lack of moisture. This animal has been reared through a number of generations as stock entered the hobby

Phyllovates chlorophaea Texas unicorn mantis.

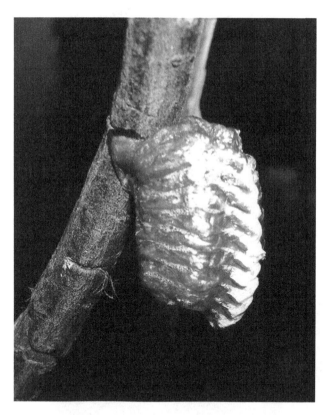

Phyllovates chlorophaea ootheca.

from a single source, from a single female from Brownsville, Texas, in 2007. However, some of the present stock came from females collected in Estero Llano Grande State Parks in 2010. The Texas unicorn mantis is not a difficult species, but a number of males and females should be kept to ensure future generations because successful mating is a challenge. While nymphs rarely bother each other, males are less often safe from females, so they are not kept together except for mating. Successful copulation is more likely if there's more than one male in the cage and the ambient temperature is 80°-85° F. Even the oothecae are identical to the following species and usually hatch after thirty-five days (Y. Saw reports 38-55 days for the Estero Llano stock). As many as ten oothecae can be created, though four or five good oothecae is not a bad result. They average fifty hatchlings per ootheca. Eggs do not diapause and would die if kept too cool. All captive stock of this species originated from Brownsville, Texas. However, the species ranges south to northern South America so it is sometimes listed as the Mexican, Central, or South American unicorn mantis.

Pseudovates arizonae Arizona Unicorn Mantis
The Arizona unicorn mantis is the most beautiful and exotic looking of the U.S. mantids. Its coloration is similar to the Texas unicorn mantis but the colors are brighter and the legs and abdomen are lobed. The adult tegmina are brighter green with similar brown spots. A large horn composed of two sections rises from the center of the head between the ocelli. Nymphs of *P. arizonae* are good feeders, but are not very aggressive towards one another. The time spent between hatching and adulthood can take over a year in captivity. Males and females are similar in size and appearance, but females tend to grow more quickly than males. Stock of this species has entered the hobby at least four times since the late 1990s, but they are always lost in a generation or two. 100% survival of nymphs is surprisingly usual, but males can mature long after females and rarely show interest in mating. High temperatures may improve the male's libido. The ootheca is globular and one half to three-quarters of an inch in diameter. Oothecae contain forty to sixty eggs that hatch in thirty to forty-five days at room temperature and should never be refrigerated. In the wild there is still only one generation per year because it overwinters as small nymphs. *Pseudovates* nymphs have been collected in January and February in Arizona, indicating oothecae do not diapause in nature (McMonigle 2011). *Pseudovates arizonae* is found only in Arizona, though some sources list it from adjacent Mexico despite lack of evidence.

Stagmomantis californica California Mantis
This species is overall very similar to the following, common species, but the female's

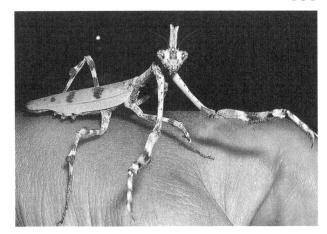
Pseudovates arizonae adult female.

hindwings are yellow along the top and the stigma is mostly white. In *S. carolina* females the stigma is dark and the hindwings are mostly opaque black. Also, when the wings are lifted, four dark bands can be seen across the dorsal surface of the abdomen. Oothecae contain 20-60 eggs and look similar to the following species except the sides usually have whitish patches. Unlike *Tenodera* spp. and *Mantis religiosa* oothecae that are spongy and squeezable, these are pretty solid. It is found in California, but also east to Colorado, New Mexico, and Texas.

Stagmomantis carolina Carolina Mantis
The Carolina mantis grows up to two and a quarter inches long and is our most common, widespread, and variable native species. It is the state insect of only South Carolina but the common name could certainly be a limiting factor. Specimens that are not green are usually mottled like gray bark but they can be brown, tan, yellow, nearly pink, or black. The variable adult color has led to at least 43 different scientific names having been mistakenly attributed to this species by novice and expert entomologists (Thomann 2002). The hind wings are dark, nearly black, but there is orange-brown near the top and base. Males possess very dark

Pseudovates arizonae nymph.

Stagmomantis californica female hindwing color.

Stagmomantis californica female stigma color.

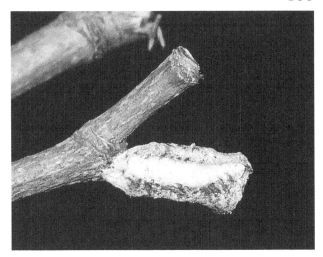

Stagmomantis carolina ootheca.

Stagmomantis carolina green female color.

wings which are sometimes a deep cobalt blue. *Stagmomantis carolina* is very easy to keep and has few molting problems. If young are kept together for the first few weeks the losses due to cannibalism are staggering. They attack each other head-first and it does not take long for the entire cage bottom to be covered in headless nymphs. Adult females are voracious, but they usually will not eat males unless the keeper puts no effort in the courtship process (such as throwing them both in a small cup). Females often form oothecae on thick twigs where they are not very noticeable because they can be mistaken for an old wound or gall. I have also found a number of oothecae on brick buildings four to twelve inches above ground level to as high as the top rung of a wooden fence. Nymphs hatch out after about a month, or oothecae can be stored in the refrigerator for hatching at a later time. This *Stagmomantis* species is found from New Jersey west to Utah and south all the way to Central America.

Stagmomantis limbata Bordered Mantis
This is the bulkiest of the native *Stagmomantis* and it seems to be the most commonly encountered species in many areas of the southwest. The name "bordered" comes from the narrow, light colored border on the green or brown tegmina. The most conspicuous difference from *S. californica* and *S. carolina* beyond size is the female's canary-yellow hindwing coloration which sports a checkerboard of clear windows. Oothecae are rarely less than an inch long and contain up to a hundred eggs. Hatching of all nymphs from an ootheca often takes place over a twenty-four hour period. Overall it is an easy to keep species, but the nymphs are nearly as

Stagmomantis carolina ash female color.

Stagmomantis limbata brown female.

KEEPING THE PRAYING MANTIS

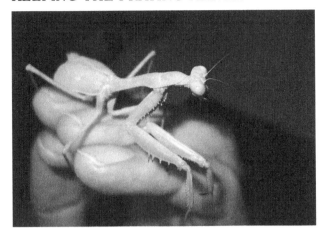

Stagmomantis limbata green female.

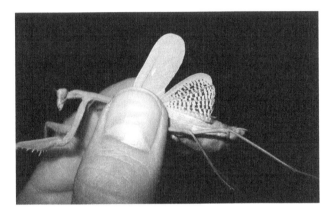

Stagmomantis limbata female hindwing color.

cannibalistic as *S. carolina*. It is common in Arizona and ranges to California, Texas, and Mexico.

Stagmomantis floridensis Larger Florida Mantis
The larger Florida mantis is the longest of our native *Stagmomantis* and females can exceed three and a half inches. Even when it is longer, the abdomen is notably thinner and less massive than the rotund *S. limbata*. The female's hindwings appear similar to *S. limbata* with a yellow checkered pattern, but the wings and tegmina are shorter in relation to the body. One easy way to distinguish it from *S. carolina* from the same areas is the stigma, which is mostly white. This species is found only in Florida.

Stagmomantis floridensis nymph. © Richard Trone

Tenodera angustipennis Narrow-winged Mantis
This is usually considered our least common adventive, but it looks so similar to the next species it may be more common than it seems. *Tenodera angustipennis* is a slim and good-sized species usually over three inches in length. There is a bright orange spot between the forelegs, which can be seen on adults and large nymphs. The hindwings are see-through and tinted a pale, pinkish-tan. Females form an elongated, flattened ootheca with a pair of longitudinal grooves that cannot be confused with other mantids. Overwintered oothecae

Tenodera angustipennis wing color.
© Richard Trone

Stagmomantis floridensis adult female. © Tony DiTerlizzi

Tenodera sinensis wing color.

Stagmomantis floridensis and *Stagmomantis carolina* adult female comparison. © Yen Saw

Tenodera angustipennis ootheca (left) and *Tenodera sinensis* ootheca (right). This *T. sinensis* hatched out hundreds of nymphs in a few hours while the *T. angustipennis* hatched out a few a day for more than a month.

Tenodera sinensis. (CC) Matt Tillett

brought indoors normally hatch out completely in a few hours but I acquired one in early winter that hatched one to five nymphs every day for thirty-five days beginning in late November. The first nymph to hatch ate one of the two that hatched the following day, leaving behind only the ends of the legs. Nymphs can be reared together when kept supplied with fruit flies but early separation is required for high survival. In contrast to *T. sinensis*, oothecae and animals are usually found at head level or higher in nature (Rick Trone pers. comm.). The narrow-winged mantis may be found anywhere in the U.S., but is normally reported from Pennsylvania south to Georgia (specimens in photos are from North Carolina). Its native range is Eastern Asia (Ehrmann 2002).

Tenodera sinensis Chinese Mantis
Our most massive species can exceed four inches and is the most commonly encountered mantis in the eastern and central United States. It can be rapidly discerned from *M. religiosa*, which is nearly as common, because the face is marked with colorful, vertical ridges. It can be discerned from *T. angustipennis* by the blackish hindwings of the adult or the pale yellow spot on the underside of the prothorax between the coxae of the forelegs when nymphs. The outer edge of the tegmina, known as the costal field, is usually green even if the rest of the animal is solid brown. Males and females are capable flyers though females commonly eat too much and grow too heavy to fly. Oothecae are the familiar globular eggcases that look like clumps of spongy, insulation foam pasted to upright sticks or tree branches four to five feet off the ground (very rarely as high up as *T. angustipennis*). A female deposits three to six oothecae, each containing 50-400 eggs (Stefferud 1952). The nymphs are rather fragile and deaths from molting accidents are high.

Tenodera sinensis green.

Some nymphs have weak abdomens that bend backwards when they hang upside-down and these may not survive to adulthood even if a special cage is built to cope with this problem. This species may show up in any state. Originally, it is from Eastern Asia. In Southeast Asia, it is replaced by *Tenodera aridifolia* (Schwarz, pers. comm.).

Thesprotia graminis Grass Mantis
This dainty creature is often found in fields or near pine trees. It is well camouflaged and barely noticeable in its habitat. The male has huge wings but they are folded tightly to maintain the slim outline. Females have no wings and look very much like a young *Brunneria*, though they are unrelated. They can be kept communally with essentially zero cannibalism.

Thesprotia graminis large nymph.

Thesprotia graminis male. © Tony DiTerlizzi

KEEPING THE PRAYING MANTIS

Thesprotia graminis oothecae. Specimens provided by Adrienne Siebert.

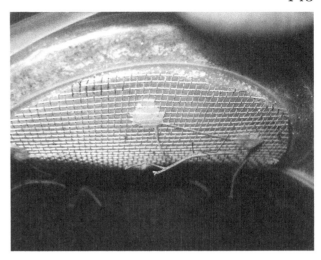

Thesprotia graminis communal nymphs.

Nymphs prefer high humidity, grow quickly, and feed on small fruit flies from the start. As they grow bigger they never seem to outgrow fruit flies. Though adults are many times larger, they are seldom in the mood to eat bigger food. Females grow to a little over two inches but the forearms are skinny and frail. There is one species ranging from Georgia to Texas and south.

Yersiniops sophronicum Yersin's Ground Mantis
Yersiniops ground mantids are generally inconspicuous but have large conical eyes. This energetic and inquisitive species is tan to brown in color and tops out at around half an inch. *Yersiniops* are also known as grasshopper mantids because they have long rear legs and can jump. They are swift runners and are seldom found above ground level. Adults and nymphs eat relatively large prey but do not do well on fruit flies. The related horned ground mantis *Yersiniops solitarium* appears similar but with less closely set eyes, and can reach nearly an inch in length. Both *Yersiniops* spp. are found in the Southwest, although the range of *Y. solitarium* does not extend into California.

Yersiniops from Arizona.

Yersiniops sophronicum. © Yen Saw

EXOTIC MANTIDS

There is an incredible variety of bizarre and beautiful exotic mantids. Many possess the standard mantis appearance: triangular head, long straight body and pronotum, and green to brown coloration. However, many exotic mantids deviate substantially from the ordinary appearance. Some are shaped like dried-up, crumpled leaves. Deadleaf mantids may have ragged extensions on the legs and thorax. Their tegmina are often exact replicas of curled or torn leaves, complete with veins and imperfections. Others look like flowers, and have bold and elaborate colors and patterns on their wings and bodies. Some have flower petal-like extensions on their legs, and can be pink, white, yellow, or almost any color imaginable. The heads come in every shape and can have strange leafy projections or intricate horns on top. Some are very long and thin and resemble U.S. grass mimics or look like a twig, complete with rough, bark-like texture. Others are broad and flat, disappearing in perfect camouflage on a tree trunk. There is a nearly endless array of strange looking mantids found throughout the world's rainforests, tropical grasslands, deserts, and transitional habitats. There is even a species found nearly 16,500 feet high along the flanks of the Himalayas (Stefferud 1952).

The continents where the most hobbyists occur (Europe and North America) each only host about two dozen species in the wild. Although there are approximately 2,450 mantis species worldwide, only a relatively small number of species have been successfully kept in captivity for multiple generations. Most have never been in the hobby. Some species do not do well in captivity, dropping dead for no apparent reason, needing very specialized conditions to be kept alive, or nearly impossible to mate successfully. Some are redundant: there is no point in concurrently maintaining ten different species from the same genus (or even similar genera) that look nearly identical. Others do well for a few generations and stop breeding, but the majority have been lost only because people stopped keeping them. Nevertheless, there are a number of exotic species that have been bred successfully for a dozen or more generations. There are only a few species where some stocks might hail back to those from the 1990s: *Phyllocrania paradoxa*, *Sphodromantis lineola*, *Hierodula membranacea*, and *Miomantis paykullii*. Included below are various tropical species that have been kept over the years and depending on when you read this, some may be old hat or new again. A number of common species are found in numerous countries but have been given common names according to the country the original stock was found in or just for the purpose of differentiation. Country of origin not expressly cited is taken from known captive stock locale and may

Acanthops sp. large nymph.

Acanthops sp., females often expose the brightly colored abdomen.

be amended with type specimen localities listed by (Otte et al. 2012) except those conflicting with Ehrmann (2002). Feeding and egg laying sites can severely affect the normal size and shape of oothecae. Average ootheca size and shape listed are those normally reported.

Acanthops falcata Venezuelan Deadleaf Mantis
Acanthops are peculiar mantids. The raptorial front legs are perfectly formed to fold up between the head and prothorax to form a scoop. The front legs comprise nearly half of the nymph's mass and young easily tackle prey larger than themselves. Males and females can reach adulthood in two months. Males live another two or three months while females continue on for six or more. Adult males have widened tegmina, many times the width of the abdomen and look nothing like the female. They are excellent flyers and skittish, which usually makes them adept at avoiding greedy females. The two-inch female holds her abdomen arched up over the body and has small tegmina, each with a whirled tail at the end. When threatened, she moves her wings to expose purple, black, and yellow markings on the abdomen. The stick-like oothecae are usually 2-3 inches long, but are not cemented lengthwise to an object. The peculiar oothecae hang from a thread. Beyond Venezuela and northern South America, it occurs north to Mexico.

Acromantis spp.
These small hymenopodids look more like a standard mantis and have no special adaptations to mimic flowers. Males are usually just under an inch and females are around an inch and a quarter. They play dead but jump around like crazy if you touch them. Oothecae are tiny and contain twelve to twenty eggs. They take from thirty-five to more than fifty days to

Acromantis sp. 'magna.' © Yen Saw

Acromantis formosana. © Yen Saw

Female *Blepharopsis mendica.*

hatch. Twenty-one *Acromantis* species are found across Southeast Asia and at least three different species have shown up in culture.

Alalomantis muta Cameroon Mantis
Alalomantis muta somewhat resembles *Sphodromantis* but with brighter markings. It had been a somewhat commonly encountered species for years. The inside of each front femur is decorated with two large white circles surrounded by black. Adults are about three inches long. Males live three or four months while females are among the longest-lived species, with a manticulturist reporting twelve months for a female (A. Yelich pers. comm.). The adults and nymphs are ravenous feeders. Oothecae are large and also resemble *Sphodromantis*. The Cameroon mantis is native to Cameroon but is found all the way east to Uganda.

Archimantis sp. Australian Giant Mantis
This is a nice, big species which is most notable for its long, flat cerci. They grow up to four inches in length but they were touted as a monstrous mantis reaching five inches. As with a number of species, they are much smaller in captivity, when a ruler is used to measure them (people tend to exaggerate). Oothecae contain a hundred or more eggs. There are nine species of *Archimantis* found across Australia (Milledge 1997).

Blepharopsis mendica Lesser Devil's Flower Mantis
This species stole its name from the devil's flower mantis through the "mistake" of importers selling it as *Idolomantis* years before the real creature showed up. It was obviously not *I. diabolica* but the common name stuck though some try to call it the "thistle" mantis, a name unlikely to be remembered due to its blandness, though it refers to the plants that

Archimantis sp., adult female.

Subadult female *Blepharopsis mendica*.

specimens are often found on in nature. Nymphs are usually tan or brown. The beautiful adult starts out brown but changes over to a beautiful green, white, and blue pattern that looks amazing. This could be among the top prettiest species, except they do not have much of a threat display. The female and male are about the same length and grow from two to two and a half inches. As usual, the male is thinner and lighter bodied. Adult males have feathery (bipectinate) antennae like other

Male antennae, *Blepharopsis mendica*.

Large nymphs *Blepharopsis mendica*.

members of the family Empusidae. The globular ootheca is about an inch long and nearly as wide, and has a little flag of oothecal material off one side. The foamy exterior is soft and never seems to fully harden. Around fifty nymphs hatch from each. With this species it is suggested the cage sides are misted rather than direct misting of the mantids or oothecae. The beautiful *B. mendica* hails from north Africa and the Middle East.

Ceratomantis sausurii
This is a spectacular hymenopodid with a huge horn on its head and large spiny bumps on the pronotum. Like *Otomantis* the antennae perpetually oscillate and it is a little smaller than

Creobroter. Females have been successfully mated a week after maturity, though there is little data on the best number of days to wait. Oothecae have only a thin outer casing and resemble an assassin bug egg cluster complete with what looks like contrasting operculums (egg lids). They contain only about fifteen eggs each. Tinnesen (2005) reported his oothecae always hatched at 10 AM (Denmark time which would be 3 PM in Thailand), but it may relate to the day night cycle rather than time. The stocks in captivity were collected in Thailand.

Chloroharpax modesta Nigerian Flower Mantis
The Nigerian flower mantis is another hardy, handsome species with females that grow to one and a half inches at maturity. Males are of course a little smaller. The female's stigma are very notable and the large white spots with thick black borders almost look like eyespots. It is a warm savanna species and does best if misted every day or two. Specimens have been

Ceratomantis sausurii nymph. © Kenneth Tinnesen

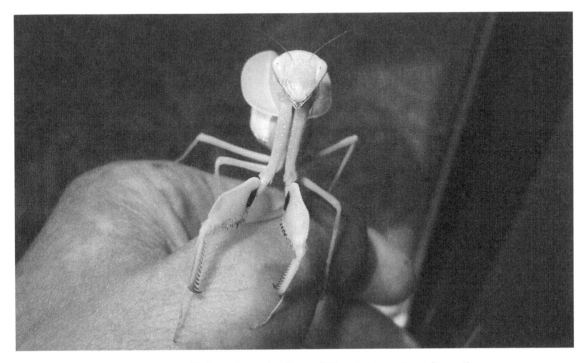

Cilnia humeralis female, note drop of blood drawn from fingertip.

Cilnia humeralis brown female nymph.
© Kenneth Tinnesen

found in Cameroon, Congo, Gabon, Ghana, Guinea, and the Ivory Coast, while the stock in captivity came from Nigeria. It is the only species in the genus.

Cilnia humeralis Wide Arm Mantis
This round-headed creature is among the most voracious and aggressive of all the species kept in captivity. The threat display is commonly employed but it has one of least colorful displays of any species. The common name comes from the unusually wide femora. Females are massive and bulky but barely reach three inches in length. Oothecae contain around two hundred eggs and are generally straightforward to hatch. It is found in southern Africa (Kaltenbach 1996).

Creobroter spp. "Flower Mantis"
Similar species in the genus include *C. apicalis*, *C. meleagris*, *C. pictipennis*, and *C. gemmatus* for a total of about twenty species found throughout tropical Asia (Ehrmann 2002). Adults take up very little space and do not need to be put into anything larger than a 16oz. deli cup. The nymphs are less likely to eat each other than many species, since they communicate by waving the eyespot on their up-turned abdomens and waving

Ceratomantis sausurii female. © Kenneth Tinnesen

motions of their forelegs, but they cannibalize if little food is around. The eyespot on the abdomen disappears on the molt to adult, and a new, colorful spot appears in the same plane on the tegmina. They mature in as little as sixty days but can take much longer if underfed. Males usually live about three months and females five, but much longer times have been reported. Females form long, thin oothecae, along tiny branches. Each usually contain about 35 to 50 individual ova.

Deroplatys desiccata Giant Deadleaf Mantis
As the common name denotes, *D. desiccata* seems to disappear among a pile of dead leaves. *Deroplatys* may not be the most extravagant

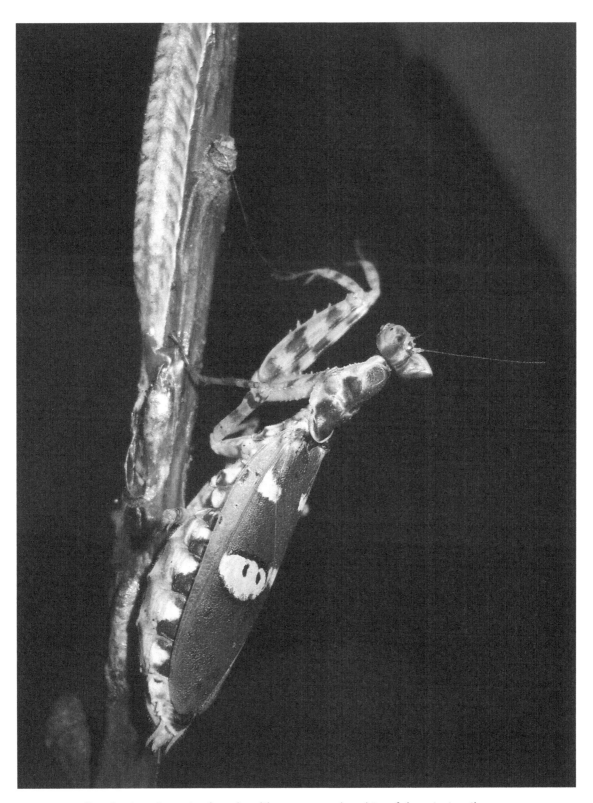
Creobroter elongatus female with one normal and two false start oothecae.

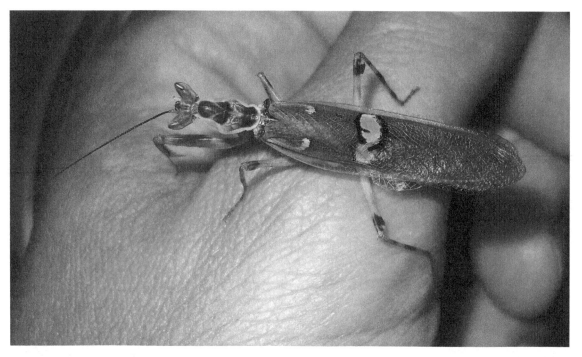

Creobroter sp. male.

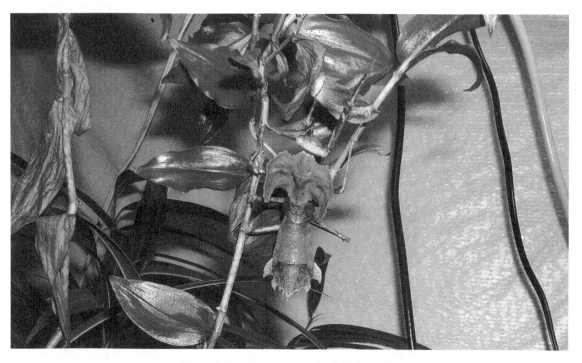

Deroplatys desiccata subadult female.

Creobroter urbanus. © Yen Saw

Deroplatys lobata female and ootheca.

of the deadleaf mimics, but they are the biggest and possess the most intimidating threat displays. The female's pronotum is about one and a half times longer than wide with downward pointed sides, while the male's pronotum narrows towards the back. The sexual dimorphism is difficult to identify on hatchlings but differentiates as they grow. Nymphs are not sexable until third instar, following the second molt. Adults are bulky and can exceed three inches in length. This species grows much larger than other *Deroplatys* that have been kept thus far. Males are nearly as long as females but are less massive. Hatchlings are huge and take the large *D. hydei* or L1 *Schultesia* readily, but they are not as aggressive as a much tinier *Hierodula* first instar and rarely cannibalize when reared communally. Nymphs often hold the body in a contorted fashion that allows them to disguise any resemblance to an insect. Deadleaf mantids are hearty feeders but rarely attack others of the same size. The adult molt can be dangerous for the female if the container is small or there are sticks in the way. Getting fertile ootheca is the only real challenge with this species. Mating is most readily accomplished on a large, inverted, horizontal screen surface and it helps to put in three or four males. The females seldom eat the males but the problem is males rarely show interest and hand mating is realistically impossible. I was able to breed seven consecutive generations (and produce many hundreds surplus) mating in this fashion and keeping them at 70°-74° F. However, the mating terrarium included a 60-watt incandescent light directly over one spot of the screen lid and the added heat may have aided mating. The light timer was set for 14-16 hours of light per day. Females live up to a year following maturity and can be alive when their offspring mature but rarely form good oothecae after the first five months. Each female normally produces three to five stocky oothecae an inch wide by three-quarters of an inch to an inch long. They are hidden among large dead leaves or crevices in tree bark. Oothecae

Deroplatys desiccata male.

Deroplatys desiccata adult female.
© Kenneth Tinnesen

contain 20-40 eggs and hatch in approximately six weeks (McMonigle 2004a). The gestation period varies greatly under similar conditions, so do not give up on an ootheca until it has been at least two full months. Found in Southeast Asia, including Borneo, Java, W-Malaysia, and Sumatra (Bischoff et al. 2001).

Deroplatys lobata Deadleaf Mantis
The female *Deroplatys lobata*'s pronotum is diamond-shaped and nearly as wide as long, while the male adult has a narrower pronotum and looks more like a standard mantis. The oothecae can be very long and contain up to 100 eggs. In contrast to the preceding species, *D. lobata* and related species deposit the oothecae on thin branches. Nymphs hatch out much smaller than *D. desiccata*. Due to a less fragile demeanor and smaller size, bad molts are rarely reported. This species has been the most common in recent years and is commonly imported directly from Malaysia to U.S. insect zoos. I took one photo at Insect World and another at the Smithsonian. There are around a dozen species in the genus that are easily determined by the shape of the pronotum. A similar but slightly smaller *Deroplatys* species available in the hobby, particularly in Europe, is *D. trigonodera*. The *D. angustata* pictured is 2004 Malaysian stock kept for a few generations. In all species, the male's

Deroplatys trigonodera. © Tom Larsen

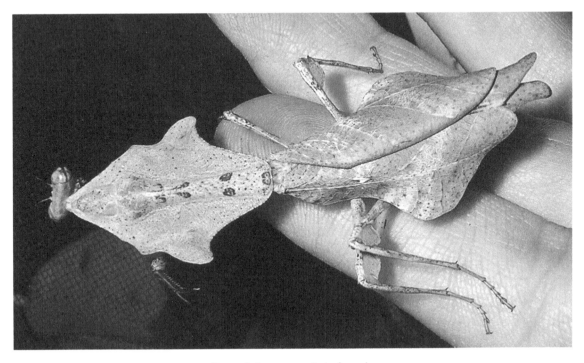
Deroplatys angustata female.

pronotum is markedly less pronounced and usually a different shape. A fourth species, *Deroplatys truncata*, has shown up a few times over the last few decades but had always been extremely rare until recently. The pronotum of the female *D. truncata* is wider than long and crescent shaped. Although captive bred stocks are usually from mainland Malaysia, the species also occur on Sumatra, Java, and Borneo (Ehrmann 2002; Delfosse 2009).

Empusa fasciata
This spectacular mantis is found into southern Europe but looks a lot like the tropical *Gongylus*. As usual for this family it cannot climb glass and can be kept communally since it is not attracted to slow moving or large prey, although mature females may cannibalize overactive males. They prefer flying prey over other food items, but will also eat an additional firebrat or roach. The female creates a dozen or more small oothecae that contain up to a few dozen eggs. The most amazing trait of *Empusa fasciata* is the oothecae commonly hatch in just two weeks. It is found from southeastern Europe to the Middle East. Similar species include *Empusa pennata* from southwestern Europe and north Africa, and *E. hedenborgii* from Egypt and the Arabian Peninsula. Other species live in tropical Africa and India (Roy 2004). The Palearctic species have a reversed cycle compared to other mantids: they hatch in July to September, grow to subadult until the onset of the winter, and overwinter in this

Deroplatys lobata male.

Deroplatys truncata pair. © Yen Saw

stage to become adult the next spring (Kaltenbach 1963). This trait renders a continuous breeding very complicated, since subadults adjust their diapause to the photoperiod and will invariable stop growing in fall unless kept in a windowless room at a constant twelve hour light cycle. Diapausing nymphs need 34-50° F and misting twice a week, but virtually no food. If kept above 59° F, they will mature prematurely and intermittently through the winter months. Diapause is to be ended in spring by slowly raising the temperatures in the course of two weeks. During late spring and summer, temperatures of 95°-104° F are required (Schwarz, pers. comm.).

Ephestiasula pictipes Purple Boxer Mantis
This is yet another of the fascinating, small hymenopodids known as boxers that has been popular in recent years. The inner femora are marked with a black-white pattern used in signaling conspecifics. They can be kept communally with some care but older nymphs and adults will eat each other. Like other boxers, they are quite tiny and despite this, the huge appetite and massive raptorial legs allow them to accept common prey. Oothecae are tiny and rarely contain more than a few dozen nymphs but are laid in quantity and hatch well. The name purple is a little imaginative but the nymphs often have purple or mauve highlights. Adults are usually gray with green wings. The purple boxer hails from India.

Eremiaphila spp.
Representatives of this genus, commonly known as pebble mantids, are difficult to rear and breed. Females stand on their back legs and hold up tiny, round wings in the threat display. Unlike most ground mantids, males also have shortened wings. Nymphs molt on the ground and do not need to hang during the molt

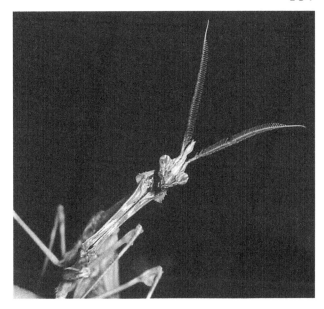
Empusa fasciata. © Yen Saw

Empusa fasciata nymphs. © Yen Saw

like other mantids. This species should be kept at 90° F or higher during the day. They run after prey at top speed and need to be kept in a round container so they can run in circles, which helps to prevent eye damage. Nymphs should be fed small roaches or *Thermobia domestica*,

Ephestiasula pictipes, hatchling "purple" boxer mantis. © Tammy Wolfe

Ephestiasula pictipes, subadult "purple" boxer mantis. © Tammy Wolfe

Eremiaphila sp. adult. © Kenneth Tinnesen

not flies. Oothecae are laid on the ground. Various members of this genus are found across the Middle East, North Africa, and Northwest India.

Gonatista sp. ex. Puerto Rico "Lichen Mantis" The adult of this fascinating mantis is two inches long and very energetic. The lichen or crab mantis runs sideways with incredible speed like a Sally-light-foot crab, quite atypical behavior for the standard mantis. Another odd characteristic is that this mantis holds its front legs off to the side rather than forward, in front of the pronotum. In nature the adults and nymphs spend their entire lives on the bark of trees and evade predators by quickly running to the opposite side of the tree. The round, flattened ootheca hatches twenty to thirty nymphs after a month. These were reared on fruit flies, then *Nauphoeta* and crickets. *Liturgusa* species look similar but the oothecae are rounded and have a spine from which the nymphs emerge (similar to *Brunneria*). There are twenty or more species of lichen mantids found throughout the subtropical and tropical Americas and Southern Caribbean Islands (Ehrmann 2002). Two unidentified species from Ecuador and Isla Margarita, Venezuela, were in culture in Europe for maybe a few generations, but they were rather sensitive and required intensive care. In nature, *Liturgusa* sp. feed mainly on ants (Schwarz 2003), but in captivity firebrats, later fruit flies and roaches, were readily taken. Freshly hatched nymphs are tiny and may not tackle fruit flies. Freshly hatched firebrats and springtails are needed here. They require thick vertical branches to live on—they can be kept communally in the first instars, but later instars and adults should each have their own stem (at least two inches wide). In contrast to some other bark mantids, they do not like to live on flat bark, but prefer a single stem in the middle of the enclosure. Accordingly, much space is needed to establish a healthy colony. Males mate readily and often, and one should separate the sexes after two or three pairings, otherwise the female may suffer from abdominal injuries (Schwarz, pers. comm.).

Gonatista sp. from Puerto Rico. © Kelly Swift

Gongylus gongylodes Wandering Violin Mantis *Gongylus* are very unusual looking creatures. A look at a mature female from above explains the fanciful common name. The pronotum is incredibly thin and elongated, while flared greatly at the apex. The thin legs possess expanded lobes and the abdomen is flared and lobed. Nymphs grow slowly and are prone to mismolts if kept below 75° F, but they can be

Gongylus gongylodes hatching. © Henry Kohler

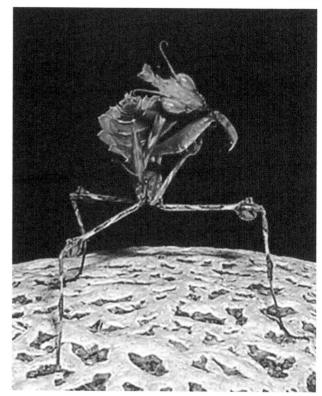

Gongylus gongylodes nymph.

Gongylus gongylodes subadult female.

kept together in a large heated cage with little or no cannibalism. The adults look similar to nymphs but are fully winged. Adults are one of the big mantids and females can reach four inches in length. Males are slightly smaller and have large, feathery antennae. Like other empusids they cannot climb glass. Wandering violin mantids can be difficult to feed anything but flies or moths, as they are ambush predators and prefer to starve than chase after anything. If they cannot reach it, it is not going to

Hagiotata hofmanni pair.

Hagiotata hofmanni female.

be eaten. Oothecae are case-shaped, one inch long by three-quarters of an inch wide, and covered with irregular shell-like extensions. They contain twenty-five to fifty eggs and hatch after six weeks. Larsen (2005) reports having kept this species for seven generations. They are found in India, Java, Malaysia, Myanmar, Sri Lanka, and Thailand (Roy 2004). Also from India, *Gongylus trachelophyllus* is similar but larger and with a wider pronotum. It has never been cultured so far. The Malaysian specimens currently promoted under this name do not belong to this species (Schwarz, pers. comm.)

Hagiotata hofmanni Paraguayan Stick Mantis
This is a handsome species though it is not really a convincing stick mimic like some of the Old World stick mantids. It is closely related to our *Phyllovates* and *Pseudovates* and is likewise a hardy species that does not usually cannibalize. However, in the same way, males and females can be kept in the same cage for months without mating. This species hails from Brazil and Paraguay.

Hestiasula brunneriana pair and ootheca. © Yen Saw

Hestiasula brunneriana Horned Boxer Mantis
At barely an inch, this is one of the spectacular, miniature hymenopodids that wave enlarged front femora to communicate their presence to other mantids of the same species. Movements of the colorful inner femur surface are, of course, also used in the courtship routine. This was one of the first boxer species kept in captivity. Oothecae hatch only around two dozen tiny nymphs that are a little smaller than *Drosophila melanogaster* (though they still readily eat them). As with many boxers, it is very hardy and can be reared to adulthood in a 5oz. cup. Specimens hail from India, listed as "east Asia" (Bischoff et al. 2001).

Heterochaeta strachani Giant Stick Mantis
Like all the biggest mantids, *H. strachani* is elongate and thin. It looks a lot like one of our female *Megaphasma denticrus* walkingsticks

Hestiasula brunneriana nymph. © Yen Saw

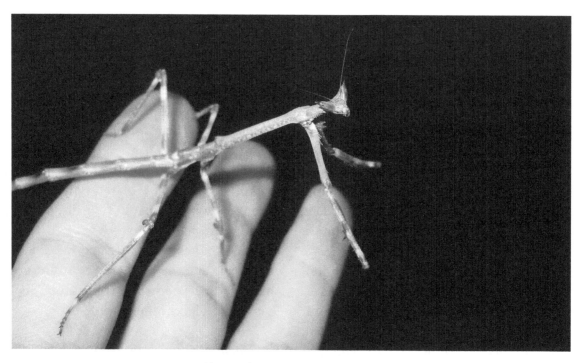

Heterochaeta sp. large nymph.

except that it has pointy eyes and develops wings. The body is knobby and very much resembles a bare branch. Females grow to a little over five inches (Lasebny & McMonigle 2001) whereas the longest stick mantis species in nature are supposed to reach somewhere around six and a half. The oothecae are elongate, dark brown, and foamy, and usually have a long, thin string of oothecal material hanging off one end. They hatch 30-50 nymphs (reported by Y. Saw). Nymphs are very easy to keep and start on fruit flies. I reared them on cockroaches and crickets, but flies and moths grab their attention better. The front legs are narrow so chosen prey should be of moderate size—they will not take full-grown crickets till two molts before adulthood. The lid should be screen and the screened area should be at least an inch longer and wider than the animal's body to provide an appropriate area for molting. I acquired three half-inch-long nymphs somewhere around 2000 but the males grew too quickly for the female despite my best efforts at slowing them down. I tried to mate her with the surviving male in a 25-gallon screen terrarium a few hours after her molt and initial mating seemed to be going well so I gave the pair privacy. I would have waited a week or two but the male was expected to die of old age any day, as his brother had, and he had been refusing all manner of prey for a week. When I returned he had changed his mind about mating and feeding. He had reversed positions and eaten a section of her teneral abdomen. No *H. strachani* stock exists at present. It is native to west and central Africa (Roy 1977).

When talking about the longest species, it is important to keep in mind different stocks, specimens, and genders may vary greatly and measurement should entail use of a vernier caliper for anything but a rough estimate. Also, body length never includes cerci, antennae, or legs, but is the length from the front of the head to the end of the abdomen or tegmina, whichever is greater. Inclusion of the legs or other appendages may be responsible for the majority of "giant" mantids that never really existed.

Hierodula membranacea.

Heterochaeta strachani had long been, by far, the longest species of which specimens ever actually existed in the U.S. hobby. In Denmark, a 20-year manticulturist stated in his mantis book a few years back that *H. strachani* was the largest species he had seen alive (Larsen 2007). In Germany, according to Schwarz (pers. comm.) *Solygia sulcatifrons* arrived more recently than 2007 and can grow longer, while in the early 2000s *Macrodanuria elongata* was in culture briefly and females could grow to nearly six inches. Recently, two other species of the genus have been introduced into culture, *Heterochaeta occidentalis* from Namibia and *H. orientalis* from Tanzania, which might grow as large or larger.

Hierodula membranacea Giant Mantis
Hierodula usually resemble a large, pastel-colored *S. lineola*, but it is a large and varied genus. The common name for *H. membranacea*

Hierodula majuscula. © Yen Saw

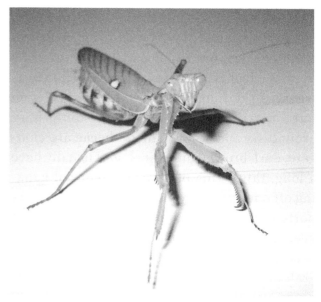

Hierodula unimaculata.

is due to its large size, but mostly results from the old myth that some of the *Hierodula* could reach six inches. Females of the similar looking *H. grandis*, the largest verifiable, grow to over four inches in length, while *H. membranacea* are rarely much longer than three and a half inches. *Hierodula grandis* have much spinier coxae and slightly different looking oothecae but have not been in culture for maybe a decade, though *H. membranacea* are sometimes mislabeled *H. grandis*. Nymphs and adults of various *Hierodula* are voracious feeders. The gigantic females can eat nearly any insect imaginable, even an adult hissing cockroach, *Gromphadorhina portentosa*, though they tend to eat through the underside and throw the top shell away. Oothecae are large and globular like *Sphodromantis*, but can be larger than a golf ball and have a papery rather than foamy exterior. My first *Hierodula* were *H. patellifera*, which are only two-thirds the size of *S. lineola* but are much prettier, with purple legs and faces and colorful nymphs. Oothecae of smaller species like *H. patellifera* tend to be solid, but large species like *H. membranacea* have a space filled with mostly air between the inner egg cluster and papery outer covering. They can become great noisemakers if one is bored. Egg counts are in the 100-300 range. There have been maybe a dozen different *Hierodula* available over the years, including *H. membranacea*, *H. patellifera*, *H. trimacula*, and *H. unimaculata*, as well as a number of unidentified species. With no less than a hundred and four species (Larsen 2007), one out of every two dozen known mantis species is in this genus. Although *Hierodula* are relatively huge mantids, none of the known species grow over four and a half inches (which is still quite impressive for such a thick-bodied mantis). *Hierodula membranacea* is found in China, India, Malaysia, Nepal, and Sri Lanka, while other species range across Asia.

Hymenopus coronatus Orchid Mantis
The orchid mantis is the most sought after and thought, by many, to be the most beautiful mantis in the world. Nymphs resemble the flowers of *Phalaenopsis* spp. moth orchids. Nymphs even have nectar guide markings like

KEEPING THE PRAYING MANTIS

Subadult female orchid mantis.

Orchid pair, note the tiny male.

the flowers, but these guides lead to a rather nasty surprise as there is no nectar. The guides are, however, on the dorsal surface of the abdomen. In the wild they are rarely seen on orchids, but are frequently found on a whole array of flowering plant species, including trees like *Melastoma polyantha* (Annandale 1900; Tomasinelli 2003; Schwarz, pers. comm.).

Hatchlings are red and black, mimicking reduviids (assassin bugs). Adults are very pretty mantids but no longer resemble moth orchids. They exchange pink and green markings for brown at maturity and males are more brown than white. The nymphs feed well but have little desire to chase after food (though they are happy to sneak up on food). Males grow very quickly and can become adults when females of the same age are two to three molts away. Adult males are very small and look like tiny flies compared to the females. Adult females live five to eight months, while adult males often live only a few months, though some live as long as seven. Well-fed adult females require good ventilation or they fall over dead (a mostly screen cage, or air circulated by a fan or aquarium air-pump is sufficient). The oothecae are stick-like, as with many hymenopodids, but they are surrounded by thicker, spongy foam and are a quarter inch across by two to four inches in length. Females produce two to five eggcases, seven to thirty days apart (McMonigle 2001). Oothecae contain twenty to a hundred eggs, but usually fall somewhere near the middle. *Hymenopus coronatus* is found in tropical Asia, including Borneo, Indonesia, Java, Malaysia, and Thailand.

Idolomantis diabolica Devil's Flower Mantis
This is the original devil's flower and until a few years ago it was an unobtainable "dream" mantis. The illustrious history and breeding requirements of this species were compiled by Schwarz et al. (2009). Though unassociated with prey capture, the threat display is what originally earned this species the devil's flower name. In 1899 Sharp imagined the mantis sat around all day in its threat display pretending to be a flower, though he had never seen a live one (and many authors copied his report). Prete et al. (1999) suggested it might become

Idolomantis adults. © Yen Saw

Idolomantis diabolica 2nd instar and 1st instar hatchling.

Cage adapted for *Idolomantis diabolica*. © Nick Jackson

environmental or dietary issues, but placement of sticks in the upper area of the terrarium and an increase in humidity helps). Screen is important since it cannot climb smooth surfaces like glass. Metal screen should be avoided as it can lead to tarsi damage, while even limited smooth surfaces can cause molting problems in later instars when they try to hold onto such. Cloth cages are recommended, but low ambient humidity can dry them out, especially if a light bulb is used to raise the temperature. Like most empusids, it does best if offered a diet primarily of flies and moths, since it is difficult to use other prey in such a way as to initiate a feeding response in an appropriate cage. Some specimens take fluttering roach males offered from tongs. It has done well in captivity for a number of consecutive generations, but it is a very difficult species. The natural range across Africa is extensive, from Cameroon to southern Ethiopia and south to northern Mozambique (Roy 2004).

Metallyticus violaceus

They are nearly black with reddish femora and metallic blue highlights that are difficult to appreciate in a photo, though adults are not

extinct from the clearing of African forests, but it is a savannah species so reckless pesticide use or modification of savannah habitats would be far more credible threats. The Devil's Flower is arguably the most spectacular combination of size, color, and shape of all the Mantodea, though it falls near the bottom of the list when it comes to attitude. The adults are among the bulkiest of all mantids, females can grow to 4.3" and have wide flaring projections and appendages. It is prone to molting trouble in the late instars and a percentage of the subadults often do not molt correctly (possibly due to

KEEPING THE PRAYING MANTIS

Metallyticus violaceus adult. © Myke Frigerio

and a half at maturity. It is very easy to breed a few generations a year. This is undoubtedly the best species for enthusiasts interested in trying their hand at reproduction. Nymphs grow quickly and are easily reared in small cages. Bad molts are rare. Males can be longer than the female, but have larger wings and a graceful build. Females will eat males, but a one-gallon breeding container is big enough for a successful mating cage. Adult males live two to five months. After maturity is reached, females live as long as nine months and make six to twelve eggcases. Oothecae are an eighth of an inch wide and vary from a quarter to three quarters of an inch long; twenty to seventy nymphs will hatch after thirty days. Other similar African *Miomantis* species include *M. caffra*, *M. abyssinica*, and *M. binotata*, but identification can be difficult as this is one of the largest genera with around seventy species

spectacularly colored like the sexually dichromatic *Metallyticus splendidus*. Adults are generally one and a half inches long or better. It grows well when fed various cockroaches but will need small, newborn *Schultesia* or young firebrats in L1. Cannibalism is rare unless crowded and underfed. This genus needs flat bark for hunting and molting. *M. violaceus* is found from India to Mindanao (Wieland 2008). *Metallyticus splendidus* is similar, but is not as tolerant to low humidity as *M. violaceus*. It ranges from India to Borneo.

Miomantis paykullii Egyptian Pygmy Mantis
Miomantis paykullii has the classic mantis look but measures in at just an inch to an inch

Miomantis paykullii subadult male.

Miomantis binotata. © Tom Larsen

Omomantis zebrata.

KEEPING THE PRAYING MANTIS

Odontomantis planiceps mating adults. © Yen Saw

(more than is found in a number of other families) (Ehrmann 2002).

Odontomantis planiceps Ant Mantis
Early instars resemble ants but later instars are green and look like a standard mantis. Adults are small, at less than an inch, and are just normal-looking green mantids. This is an easily kept beginner's species. Wild specimens are found in rainforests, forest edge habitat, and gardens in India and Southeast Asia.

Omomantis zebrata Zebra Mantis
Adults are among the most unique looking mantids, but the nymphs are mostly green and do not betray their spectacular future appearance. The zebra mantis is a medium-sized species at two and a half inches at the most. Thirty to fifty nymphs hatch from the ootheca. It ranges from Kenya to South Africa.

Orthodera novaezealandiae New Zealand Mantis
New Zealand is home to a single native mantis species. For a number of years this was listed as *O. ministralis*, a common Australian species, and was thought to be an accidental introduction (Forster & Forster 1980), though it was described a hundred years before it was publi-

Omomantis zebrata.

cized. Adults are rarely more than an inch and three quarters but they are usually bright green. The widened, rectangular pronotum gives them, nymphs especially, a unique, leaf-like look. This handsome little creature has been readily available to hobbyists a few times and seems pretty hardy but apparently is not terribly easy to keep through more than a few generations. Oothecae are half an inch by a quarter and contain at least twenty eggs each. In recent times, *Miomantis caffra* has been accidentally introduced to New Zealand (Ramsay 1984).

Orthodera novaezealandiae. © Cain Eyre

Otomantis scutigera Tanzanian Boxer Mantis
Adults can have a green or brown pronotum and tegmina though the rest of the body is always brown. Both sexes have full-length wings and the larger females are rarely over an inch long. There are a number of different, small hymenopodids commonly called boxer mantids because they have enlarged, colorful front femora used to signal other members of their own species. Nymphs stretch out one front arm at a time in signal motions when they first encounter one another. The "boxing" occurs at a distance and the behavior can be seen in other situations including solitary conditions. The unusual femora resemble boxing gloves on a skinny arm and the behavior gives the impression of shadow boxing. Adults are less prone to the behavior and the enlarged femora no longer dominate the body dimensions. Although the oothecae are not especially small (an eighth by a quarter inch), the hatchlings are tiny. Large springtails and small fruit flies are good prey for the first few instars. It is surprising the miniature hatchlings readily eat fruit flies since they are the same size. This species is easily kept through at least ten generations. Each female produces around half a dozen oothecae. Each of those contains twenty to thirty eggs that hatch after thirty days.

Oxyopsis gracilis
This species is sometimes called the South American green mantis but it is not always

Otomantis scutigera adult female.

Otomantis scutigera adult male.

KEEPING THE PRAYING MANTIS

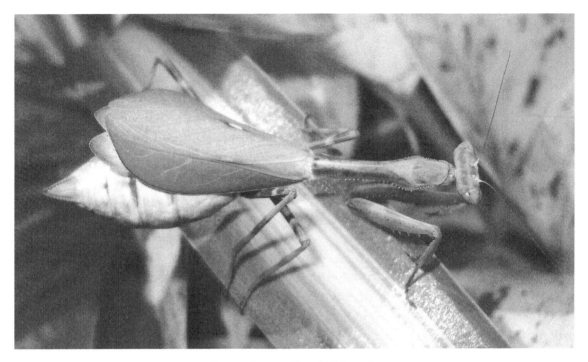

Oxyopsis gracilis adult female.

Parymenopus davisoni young adult female. The pronotum fades to yellow a few days following the ultimate molt. Leg petals can change to yellow or retain rose.

Oxypilus pallidus L3. © Yen Saw

green and there are a lot of mantids from South America. The generic name refers to the acute (pointed) eyes, while the species name means graceful or thin. The males are very thin, though the females are thick and hearty feeders. Females are just under two inches and males average one and a half inches. Found in Bolivia and Paraguay.

Oxypilus pallidus

This is another of the many small species commonly known as boxer mantids due to the oversized femora. The flightless females barely reach 0.75" at adulthood while the males fly. Like pretty much all the boxers it feeds on *D. melanogaster* at L1 and can be reared to maturity in 5oz. cups. It comes from savannas in Gambia, Senegal, and Sierra Leone.

Parasphendale affinis African Banded Mantis

Banded mantids were once among the more exotic-looking of the commonly available mantids since they are easy to reproduce in bulk. The curved, raptorial front legs are banded in light and dark and the jagged pronotum has a row of teeth that run down each side. Adult females grow to about two and a half inches while the smaller males are less than one and a half inches in length. The underside of the tegmina of female *P. affinis* is solid black. Adults and nymphs are excellent and aggressive feeders though the tiny adult males are unlikely to take out a big cricket. Female banded mantids produce three to six spongy, globular oothecae that are 1.25" by 1.25-2" long. Each contains from one hundred to two hundred eggs. This was a popular species in the mid- to late '90s but I have not seen one since 2001.

Parasphendale agrionina Kenyan Banded Mantis

This mantis is also familiarized as "budwing" which could be used to describe all the short-winged mantids. Specimens are similar to, but about 20% larger and more colorful than, *P.*

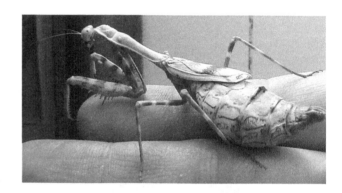

Parasphendale agrionina.

Parasphendale affinis adult pair.

Small *Parymenopus* nymph.

affinis. The base color is gray (wild-caught females are often neon green) with various splotches of green, pink, and purple. It has long been a popular species, and supplanted *P. affinis* in captive culture. Oothecae are darker than those of *P. affinis* but are similar in size and texture. Each likewise contain one to two hundred eggs. It is easy to differentiate from the above species because there is a large yellow band at the middle and end of the underside of the female tegmina that roughly matches the dorsal pattern. *Parasphendale* are found across the African savannas, though the common imports in the past were all from Kenya and Tanzania.

Parymenopus davisoni Yellow Orchid Mantis
If the orchid mantis is the most sought after mantis in the world, this species is its spectacular, ultra-rare, alter ego. The nymphs look very similar but they are often greenish white and do not grow as large. If placed on red or purple decorations they can start to take on a pink color within a few days and continue to darken over subsequent weeks. Even if they are pink it is easy to tell them from *Hymenopus* because the margin at the base of the pronotum is brown rather than green, the leg petals are smaller, and the "nectar guides" are missing. They start out much smaller than *Hymenopus* and are far more prone to death and bad molts from low humidity early on. (After the first few molts they become hardy and eat *Nauphoeta* like a champ.) The relatively miniature male reaches maturity in two and a half months while the massive female takes three from hatchling to the ultimate molt. Though the nymphs look similar to *Hymenopus* and even start out with the orange and black pattern as hatchings, the adult female looks a bit different and is only about two inches. Her tegmina are much more rounded than a female orchid mantis, with three brown splotches on each tegmen. She is an overall brilliant lemon-yellow color but can retain pink shading on the leg lobes. Oothecae are long, thin, and light tan, resembling a small *Hymenopus* eggcase. Captive stock comes from Malaysia, while the species ranges over all Southeast Asia (Ehrmann 2002).

Phyllocrania paradoxa Ghost Mantis
The genus name means "leaf head" and the head is outfitted with a twisted, leaf-like projection on top. The walking legs also have smaller, leaf-like lobes and the abdomen has a pair of flat, ragged extensions on each side. This may be the only species that is magnificent and always commands interest, but is also easy. (Other mantids tend to be either uniquely amazing to look at or easy to keep, rarely both.) The males are easily identified after two molts because the expansion of the male's prothorax is much smaller and the leaf on the head is pinched in the middle. Both genders are about two inches long when fully grown but the female is much more massive. Her tegmina look like dried brown leaves complete with veins, while the male's tegmina look more like the hindwings and give him an odd, rectangular shape. This mantis grows slowly, and lives longer than most species. *Phyllocrania paradoxa* can spend nine months or more reaching

Ghost mantis female.

Ghost mantis male.

Some variation in ghost mantis colors.

adulthood (Jiang 2006). As newly hatched nymphs, and for the first few molts, they eat and grow normally. However, as each successive molt occurs, the interval between molts gets longer and longer, with the stage prior to the final molt taking up to two months. Limited food or care could easily stretch development far past a year, but five months is normal with adequate care (pers. obs.). Ghost mantids prefer to eat food a quarter of their mass or less. They are hearty feeders but prefer to stay on a fixed perch, which should be positioned so food comes nearby. They rarely chase after prey unless it is less than an inch away (glass climbing roach nymphs are a favorite food). I raised twelve consecutive generations starting on fruit flies supplemented with phorids and followed by *Nauphoeta*. The occasional meal moth was employed to convince picky individuals to resume feeding. Once they have molted a few times they do not need much water or humidity, and only require a light misting two to three times a week. *Phyllocrania paradoxa* seldom kill each other after the third molt. For mating purposes, a ten-gallon terrarium can be used, as long as there is vegetation and places for the male to hide when mating is over. They are often kept together long-term, but females do eat males. They prefer to construct oothecae on sticks a quarter inch or less in diameter. The eggcases are similar in shape and size to *Creobroter* oothecae, but the surface is smooth and glossy and there is a long thread-like extension at one end. It ranges widely across Africa and Madagascar (Ehrmann 2002).

The body color often ranges from tan to dark brown, but can be nearly black or golden yellow. A small number of individuals are bright green. Eventual color and gender cannot be seen on the tiny black hatchlings that resemble ants (see photo). Green individuals begin to display green by the second to fourth instar. Green males change back to brown when they mature while females stay green unless kept very dry. The green is influenced primarily by genetics and humidity. Low humidity can cause a green nymph to turn brown at the next molt but high humidity will not make brown specimens green. Stock I acquired in the late 1990s originally threw only the very rare green individual. I kept only the young from the first green female—following two consecutive generations there was a green male and a number of green females. Only green males (green as nymphs, the trait is partly sex-linked since green male nymphs lose practically all green coloration following the ultimate molt) were mated to green females which resulted in approximately 80% green offspring after just three further generations (these could still be turned partly to mostly brown if kept extremely dry, though with always a tinge of green). I can reasonably conclude from the rapidly increasing percent and previous experience isolating color forms of *Acheta domesticus, Gromphadorhina portentosa, Porcellio* spp., and others, if I had not been forced to abandon the stock, after a few more generations brown individuals would have become all but nonexistent. Of course it is possible the selection process simply favored genes responsible for the propensity to develop green pigments under captive conditions rather than the ability to produce the pigments.

Pnigomantis medioconstricta Double Shield Mantis

There is only one representative of this genus though it is closely related to *Rhombodera, Hierodula*, etc. Likewise, it is a large, hardy species which is also among the more aggressive. Adults can exceed three inches in length. Unlike the various *Rhombodera* shield mantids, the expansion of the pronotum has no relevance to leaf mimicry since individuals are nearly always gray in color, rarely, if ever green. The scientific name means "constricted in the middle" so constricted shield might be a more appropriate common name. Nymphs are superb feeders at all ages and are easily started on fruit flies. At first they can be kept together for a few instars but must be separated as cannibalism becomes intense. Oothecae contain eighty to 200 eggs (Kurtz 2013). It is a relatively easy to mate species but it can be very difficult to keep females from eating males no matter how fattened up. The stock is from the island of Flores in Indonesia.

Polyspilota aeruginosa Madagascan Marbled Mantis

Specimens of the Madagascan marbled mantis are more likely to hail from Africa. Females resemble our biggest *Tenodera* but are half an inch longer and have a handsome marble pattern on the tegmina. It is a hearty species with few troubles except that it requires a large cage to match its size. Oothecae contain hundreds

Polyspilota aeruginosa, two oothecae.

Pnigomantis medioconstricta is a moderately large Asian species commonly referred to as the double shield mantis. © Tom Larsen

Polyspilota aeruginosa.

KEEPING THE PRAYING MANTIS

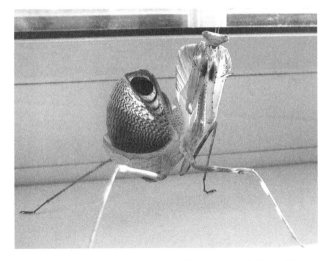

Pseudempusa pinnapavonis Peacock. © Cain Eyre

of eggs and look like a big *D. desiccata* ootheca. This species is found in Madagascar, but is also across much of Africa.

Popa spurca Twig Mantis
This thick, stocky species is an excellent mimic of a gnarled stick, complete with matching bumps and growth rings. Adults and nymphs are usually dark brown but can range in color to a light tan. The wings of the adult female are short and cover only about half of the abdomen while the male's wings are long. The tegmina are barely noticeable because they are held close and match the coloration of the body. The twig mantis feeds well and is willing to chase after large food at the bottom of the container. Compared to most grass or stick mantids the front arms are massive, which allows this mantis to eat creatures nearly its own size. Oothecae are about 3/4" long by 1/4" wide and have a soft, velvety surface. Oothecae normally contain from fifty to one hundred eggs. The deposition site depends on humidity—the drier the conditions, the closer to the ground the ootheca is formed (Schwarz 2004). *Popa spurca* is found in sub-Saharan Africa and Madagascar.

Pseudempusa pinnapavonis Peacock Mantis
The peacock mantis is named for the spectacular look of the open wings in the threat display whose markings resemble peacock feathers. This is one of the large mantids—females are heavy and can grow to three and a half inches. Large cages should be used for rearing, if possible, to reduce molting problems. Females need about six weeks to be ready to mate and seem to require plants for laying oothecae. This species hails from the rainforests of Burma and Thailand (Ehrmann 2002).

Pseudocreobotra ocellata.

Pseudocreobotra ocellata Spiny Flower Mantis
Spiny flower mantids resemble species in the genus *Creobroter*, but are much more extravagant. This genus was named after the genus *Creobroter* when that genus was known under the wrong spelling (which is why the endings do not match). The abdomen is covered with spines and the legs have lobed extensions. Older nymphs and adults are ornamented with green and pink splashes on a background of white. Tiny nymphs are blue-black with white spots. The eyespot on the back of this species is brightly colored and spiral shaped and may just serve to give their would-be attackers a headache. Adults are bulkier than most flower mantids and are around one and a half inches in length. Males and females are more similar

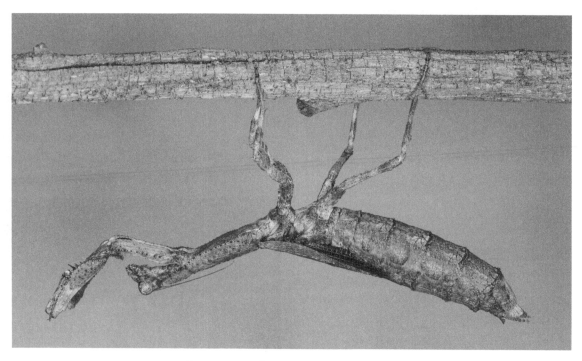

Popa spurca female. © Tammy Wolfe

Pseudoharpax virescens adult female. © Tammy Wolfe

KEEPING THE PRAYING MANTIS

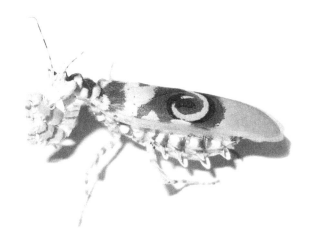

Pseudocreobotra wahlbergii.

to each other in appearance and size than many species and nymph segmentation can be very difficult to see until the later instars because color obscures the margins. Nymphs and adults seldom have trouble attacking prey that is half their mass or more. All stages are very hardy but the first couple instars desiccate rapidly while well-fed adult females require good ventilation or they fall over dead. Spiny flower

Tarachodes sp. © Tom Larsen

Pseudogalepsus nigricoxa female.

mantids are far less willing to chase after prey than *Creobroter*. Oothecae are stick like and very dark brown, nearly black. Thirty to sixty nymphs hatch after a month. *Pseudocreobotra wahlbergii*, is slightly larger, with orange-yellow rather than lemon-yellow hind wings. It occurs in eastern and southern Africa, while *P. ocellata* is from western and central Africa (Ehrmann 2002).

Pseudogalepsus nigricoxa

This hardy mantis from Tanzania grows to over two inches. Specimens look very much like *Tarachodes* and were labeled that way when

2nd instar *Pseudocreobotra wahlbergi*. © Tammy Wolfe

4th instar *Pseudocreobotra wahlbergi*. © Tammy Wolfe

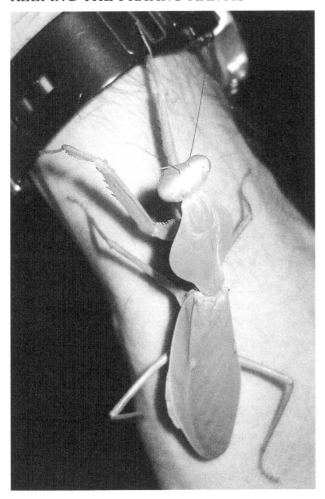

Rhombodera stalii female.

Pseudoharpax virescens Gambian Spotted-Eye Flower mantis

The common name refers to the handsome, white-spotted eyes seen on nymphs and adults. It does best with warm temperatures and daily misting. Adults grow to a little over an inch and females can produce a dozen or more oothecae that average a half-inch wide by a third of an inch long. Captive stock originated from Gambia, but the species is found elsewhere in western Africa.

Rhombodera stalii Giant Indonesian Shield Mantis

Adults are large mantids a little over three inches long with a wide prothorax. Unlike *Sphodromantis* and *Hierodula* the adults are rarely, if ever, brown. A small number of nymphs may be yellow or brown (one of each

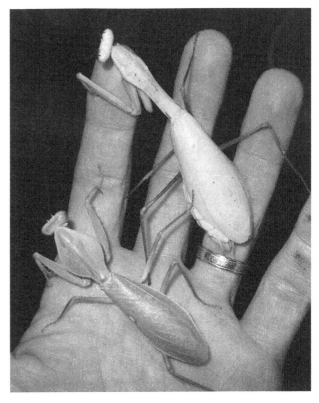

Rhombodera stalii and *Hierodula membranacea* comparison, both average females.

they first came in. These and related genera like *Tarachodula* live on stems and branches and molt more successfully if provided upright sticks instead of screen or plastic ladders for molting. The stick should be one to two times the width of the mantis and should be upgraded a few times as they grow. They feed to a high degree on ants in nature (compare Edmunds 1972) and would readily take bean weevils in captivity. In nature the flattened ootheca is often wrapped around the side of a stick and the female seems to carefully guard it from predators and parasites. In captivity she often makes it on the top or bottom of the cage. The specimens pictured were imported from Tanzania.

Brown *Rhombodera stalii* male subadult, green at maturity.

Sibylla pretiosa freshly molted adult with brown wings.

Rhombodera stalii male from brown nymph.

out of forty), but they return to green at maturity. This is a very easy species to keep and breed. Nymphs rarely experience trouble in molting though the molt to adulthood is problematic if the humidity is low and they are not misted daily (Dryer 2011). The tear-shaped oothecae are large and contain 100-300 eggs. The specimens pictured came from 2007 Java, Indonesia, stock. *Rhombodera valida* from Malaysia is another species with a larger shield

Sibylla pretiosa adult two weeks later with green wings.

KEEPING THE PRAYING MANTIS

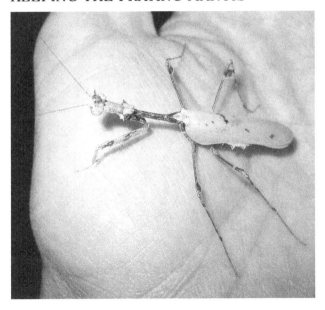

Sibylla pretiosa recently molted adult (size reference).

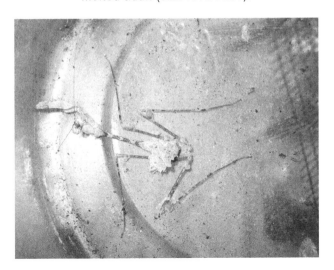

Sibylla pretiosa nymph trying to be flat.

Sibylla pretiosa yellow ootheca.

that is often kept but is more difficult to mate successfully. There are around three dozen species in the genus and some of them just look like *Hierodula*.

Sibylla pretiosa Cryptic Pretty Mantis

Despite the incredibly delicate appearance, even small nymphs are incredibly tolerant of dryness and lack of food and water when compared to most species. Adults grow to nearly two inches in length. These eat any prey but even adult females should not be given something as large as an adult cricket. Despite tolerance to desiccation they should be misted once a day and kept well-vented. Nymphs are often housed communally through the late instars, as this species has long been known to rarely exhibit cannibalism (Gale 1997), but they do well separated. Sometimes the late instar nymphs are found lying on the bottom of the cage, which for almost any other mantis would mean the animal died. They are not playing dead but are flattening themselves—it is easy to recognize this particular pose (McMonigle, 2012b). Nymphs mature in three to four months when fed *N. cinerea*, but can take twice as long if fed only flies. Mating is usually not dangerous for the male if minimal precautions are taken. The only difficulty is timing. Males commonly die about six weeks following the final molt, but can live less than four. The small oothecae are yellow and hold twenty to forty eggs each. This savanna species ranges across eastern and southern Africa. *Sibylla dolosa* is a similar looking forest species from western Africa that is sometimes available.

Sphodromantis lineola African Mantis

Members of the genus *Sphodromantis* include *S. lineola*, *S. centralis*, *S. gastrica*, and *S.*

Sibylla pretiosa nymph.

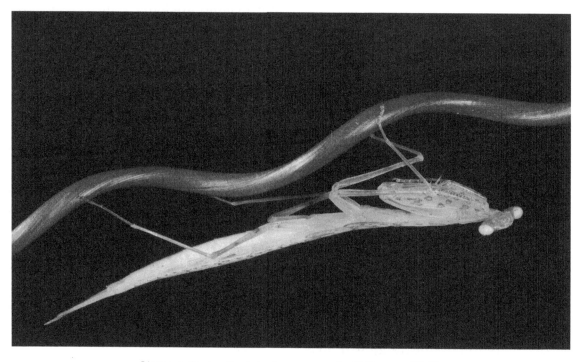

Sinomantis denticulata, large nymph. © Tammy Wolfe

KEEPING THE PRAYING MANTIS

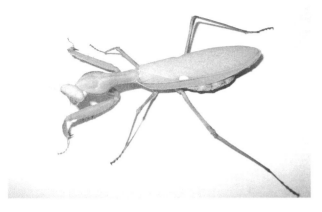

Sphodromantis viridis female.

viridis, with forty-two *Sphodromantis* species (Larsen, 2007) presently known. *Sphodromantis lineola* and *S. viridis* have been the most commonly kept in captivity, while *S. lineola* has been kept for at least ten consecutive generations. This is among the oldest of pet insects, as *S. lineola* has been kept in captivity for at least forty-four years (Stanek 1969). The various species are similar, with slight differences in markings and body structures, but

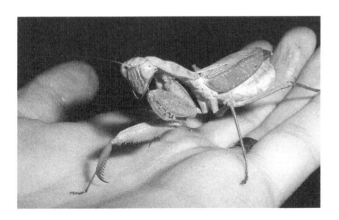

Sphodromantis lineola female.

they are all heavily built, around three inches long, and the tegmina have a notable stigma. *Sphodromantis viridis* is among the more unique with its banjo-shaped pronotum and strong blue cast to green specimens. All species kept so far have had the same rearing requirements. *Sphodromantis* are among the easiest to keep

Tarachodula pantherina male. © Kenneth Tinnesen

of all mantids and any make a great first species. The female can form up to ten oothecae, though four is more common, and each can hatch out 200 or more nymphs. It usually takes about three months for them to reach adulthood, but as with all species, reduced feeding can lead to far longer development requirements. Females are among the longer-lived adults and can live nine months. Members of this Old World genus are found in various parts of Africa and the Middle East.

Tarachodula pantherina Panther Mantis
The beautiful member of the family Tarachodidae was a dream mantis found pictured in books until it finally showed up in the hobby around 2002. Females have tiny wing buds, while males can fly. The adults are just as pretty as a picture, but the oothecae seem to always

Tarachodula pantherina. © Kenneth Tinnesen

Tarachodula pantherina nymph. © Kenneth Tinnesen

Taumantis sigiana female.

Comparison of oothecae; *Oxyopsis gracilis* (left) and *Taumantis sigiana* (right).

be grayish, unlike the spectacular purple one known from a photograph (Preston-Mafham, 1990). This is one of the strange species that only molts and hunts well on sticks that are 1/2"-3/4" in diameter. It will often sit on the floor rather than climb cage surfaces. It is from eastern Africa (Ehrmann 2002).

Taumantis sigiana
This species is most notable because it looks so much like *Oxyopsis gracilis*. Females are the same shape and size and if they are the same color it is only easy to tell them apart because this species has tiny black stigmas on the tegmina. (*O. gracilis* stigmas are larger white spots surrounded by a thick black ring.) Whereas *Oxyopsis* have little 1/2" x 1/2" dome-shaped, hardened oothecae, this species has huge, spongy, green oothecae. The 3/4" by 1.5" green oothecae hatch out bright orange nymphs that contrast greatly with the ootheca they sit on at first. Nymphs are very aggressive and difficult to cohabitate. It comes from Kenya and Tanzania.

Theopropus elegans Elegant Mantis
Adult females of the elegant mantis are larger than most flower mantids at approximately two inches. As usual for large flower mantis species, males are short and tiny compared to the females. Nymphs are excellent feeders, enjoy nearly any live food, and grow quickly. They do chase after food, but only when fairly hungry. Oothecae are up to three inches long by three-

Theopropus elegans nymph.

fifths wide and contain 40-60 eggs. The difficulty with this species is getting fertile oothecae. *Theopropus elegans* is found in tropical Southeast Asia and specimens usually come from Malaysia.

Hymenopus coronatus, hatching. © Tammy Wolfe

Idolomantis diabolica. © Kenneth Tinnesen

Hymenopus coronatus nymph. © Tammy Wolfe

GLOSSARY

ABDOMEN: The last section of the insect body, attached to the thorax; contains the genital organs and cerci.

ADVENTIVE: Non-native species which has become established; naturalized.

APTEROUS: To be without or lack wings.

ANTERIOR: At or towards the front.

APEX: Point furthest from the point of attachment.

AUTOTOMY: The reflexive release of an appendage to a specific regrowth point.

BIFURCATION: The splitting of a main body into two parts.

BIPECTINATE: Having comb-like teeth on two sides.

CERCUS (*pl.* CERCI): Paired appendages at the end of the insect's abdomen.

COSTAL FIELD: Front margin of the wing between the first two major veins.

COXA (*pl.* COXAE): Basal leg segment.

CRYPSIS: Mimicry of common background objects like rocks or plants rather than fauna.

DEIMATIC DISPLAY: The defensive posture of a mantis that often includes colors and may include sound; threat display.

DEMI-JOHN: A large, glass, narrow-necked bottle.

DESICCATE: To dry out.

DIAPAUSE: A period of inactivity or arrested development influenced by seasonal changes in temperature or humidity, such as overwintering.

DIMORPHIC: Possessing two distinct forms, shapes, or coloration. Sexually dimorphic males and females are visibly different.

DORSAL: The upper surface or top.

ECOMORPHS: Creatures whose appearance is determined by the local ecology.

EUPLANTULA (*pl.* EUPLANTULAE): Adhesive pad on the mantis tarsus used in climbing smooth surfaces; similar to pulvillus.

EXUVIUM: The shed exoskeleton.

F1: Offspring of wild-caught adults (the offspring of two F1 would be F2 and so on).

FEMORAL BRUSH: Patch of setae on the femur used to clean the compound eyes.

FEMUR (*pl.* FEMORA): Thigh; segment of an insect leg between the trochanter and tibia.

FRASS: The solid waste of mantids and other invertebrates.

INSTAR: Nymphal stages after hatching and between molts (i.e., a third instar nymph would have molted twice). Common abbreviation for instars is L1, L2, etc. The L comes from *Larvenstadium*, the German word for instar.

INTERSEGMENTAL MEMBRANE: Elastic, less sclerotized portion of the exoskeleton joining segments.

LARVA (*pl.* LARVAE): The stage between egg and pupa; usually for insects with a complete metamorphosis.

LOBE: A rounded projection of the exoskeleton.

Tarachodes sp. from Kenya. © Kenneth Tinnesen

Cilnia humeralis. © Kenneth Tinnesen

MANTICULTURIST: Person who studies the husbandry and captive life cycle of Mantodea.

MARGIN: A border or edge.

MEDIUM (*pl.* MEDIA): The food as well as substrate used in culturing fruit flies, bacteria, etc.

MESOTHORAX: Middle section of the thorax; contains the tegmina and middle legs.

METATHORAX: Terminal section of the thorax, attached to the abdomen; contains hind legs and hind wings.

MOLT: The act of shedding the exoskeleton.

NYMPH: Stage between egg and adult; for insects with an incomplete metamorphosis.

OCELLUS (*pl.* OCELLI): Light detection organs.

OMMATIDIUM (pl. OMMATIDIA): Structural unit or complete facet of the compound eye.

OOTHECA (*pl.* OOTHECAE): A group of eggs produced together and surrounded by foam or a hardened outer shell; eggcase.

PARALLAXIS: Difference in the perceived position of an object from two different lines of sight.

PARTHENOGENESIS: Reproduction through the development of unfertilized eggs.

PROFEMUR (*pl.* PROFEMORA): Third leg segment (femur) of the front pair of legs attached to the prothorax.

PRONOTUM: Shield-like upper surface of prothorax.

PROTARSUS (*pl.* PROTARSI): Fifth leg segment (tarsus) of the front pair of legs attached to the prothorax.

PROTHORAX: Front section of the thorax; attached to the head and contains front pair of legs.

PROTIBIA (*pl.* PROTIBIAE): Fourth leg segment (tibia) of the front pair of legs attached to the prothorax.

SCLEROTIZE: To cause rigidity of the exoskeleton through addition of the protein sclerotin or substance other than chitin.

SENSILLUM (*pl.* SENSILLA): An arthropod sensory organ.

SETA (*pl.* SETAE): Thin bristle or hair-like extension of the arthropod exoskeleton.

SEXUALLY DICHROMATIC: Of or having differently colored male and female forms.

SPERMATHECA: Female's sperm storage compartment.

SPERMATOPHORE: Sperm packet.

SPIRACLE: The external opening of the tracheal respiratory system; breathing hole.

STERNITE: Ventral abdominal plates.

STIGMA: The spot near the outer edge of the tegmina found on many mantis species.

TARSUS (*pl.* TARSI): The foot of an insect. The tarsus contains a number of small segments including the claws.

TEGMEN (*pl.* TEGMINA): The tough and thickened front pair of wings.

TENERAL: Soft, often pale or white colored state of the insect exoskeleton immediately following a molt.

THORAX: Middle section of the insect body; in between the head and abdomen.

TIBIA (*pl.* TIBIAE): Segment of an insect leg between the tarsus and femur.

TRACHEA: Internal breathing tube that opens at a spiracle and branches out to the body cells.

TRUNCATE: Cut off or appearing to be cut off at the tip.

VENTRAL: Bottom or underside.

VOLVATING: The property of being able to curl into a spherical shape.

Tropidomantis sp ex. Thailand. © Kenneth Tinnesen

Not all of the Hymenopodidae mimic flowers.

BIBLIOGRAPHY

Annandale, T. N. (1900) Observations on the habits and natural surroundings of insects made during the Skeat Expedition to the Malay Peninsula, 1899-1900. *Proceedings of the Zoological Society of London*, 837-869.

Arnett, R. H. (1993) *American Insects: A Handbook of the Insects of America North of Mexico*. Sandhill Crane Press, Gainesville, FL.

Berg, M., Schwarz, C. J., and J. E. Mehl. (2011) Die Gottesanbeterin, *Mantis religiosa. Die Neue Brehm-Bücherei Bd.* 656. Westarp Wissenschaften, Hohenwarsleben, Germany.

Bischoff, I., Bischoff, R., Heßler, C., and M. Meyer. (2001) *Praxis-Ratgeber. Mantiden–Faszinierende Lauerjäger*. Chimaira, Frankfurt, Germany.

Bragg, P. E. (1996) Mantids and Cockroaches Meeting and Study Group. *The Bulletin of the Amateur Entomologists' Society*. 55(405): 56.

—. (1997) *An Introduction to Rearing Praying Mantids*. Bragg, Nottingham, U.K.

Casterena, A., and J. R. Codd. (2010) Foot morphology and substrate adhesion in the Madagascan hissing cockroach, *Gromphadorhina portentosa. Journal of Insect Science* 10(40): 1-11.

Crane, J. (1952) A comparative study of innate defensive behavior in Trinidad mantids (Orthoptera, Mantoidea). *Zoologica* 37(20): 259-293.

Delfosse, E. (2009) Taxonomie, biogéographie et biologie de deux espèces de mantes asiatiques du genre *Deroplatys* Westwood, 1839, et notes sur trois autres placées dans le même genre (Insecta: Mantodea: Mantidae). *Le bulletin d'Arthropoda*, 39(1): 4-28.

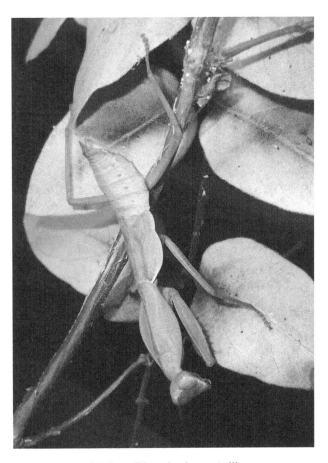

Yellow *Rhombodera stalii* female nymph (the adult is green).

Harpagomantis tricolor. © Yen Saw

Deyrup, M. (1986) Observations on *Mantoida maya* (Orthoptera: Mantidae). *Florida Entomol.* 69(2): 434-435.

Dryer, R. (2011) The shield mantis *Rhombodera stalli* Giglio-Tos, 1912. *Invertebrates-Magazine* 10(3).

Dunn, G. A. (1993) *Caring for Insect Livestock: An Insect Rearing Manual.* Young Entomologists' Society, Lansing, MI.

Edmunds, M. (1972) Defensive behaviour in Ghanaian praying mantids. *Zool. J. Linn. Soc.* 51: 1-32.

—. (1976) The defensive behaviour of Ghanaian praying mantids with a discussion of territoriality. *Zool. J. Linn. Soc.* 58: 1-37.

Edmunds, M., and D. Brunner. (1999) Ethology of Defenses against Predators. In: Prete, F. R., H. Wells, Wells, P. H., and L. E. Hurd. (1999) *The Praying Mantids.* Johns Hopkins University Press, Baltimore and London. pp. 276-299.

Ehrmann, R. (2002) *Mantodea Gottesanbeterinnen der Welt.* Natur und Tier-Verlag, Münster, Germany.

—. (2011) Mantodea from Turkey and Cyprus (Dictyoptera: Mantodea). *Articulata* 26(1): 1-42.

Farb, P. (1962) *The Insects.* Time-Life Books, New York, NY.

Forster, R. R., and L. M. Forster. (1980) *Small Land Animals of New Zealand.* John McIndoe Limited, Dunedin, New Zealand.

Frye, F. L. (1992) *Captive Invertebrates: A Guide to Their Biology and Husbandry.* Krieger Publishing Company, Malabar, FL.

Gale, M. (1992) Rearing mantids. *The Bulletin of the Amateur Entomologists' Society.* 51(382): 125-127.

—. (1997) Captive breeding of mantids. *The Bulletin of the Amateur Entomologists' Society.* 56(414): 172-174.

Gemeno, C., and J. Claramunt. (2006) Sexual approach in the praying mantid *Mantis religiosa* (L.). *Journal of Insect Behavior* 19(6): 731-740.

Gemeno, C., J. Claramunt, and J. Dasca. (2005) Nocturnal calling behavior in mantids. *Journal of Insect Behavior* 18(3): 389-403.

Gillon, Y., and R. Roy. (1968) Les Mantes de Lamto et des savanes de Côte d'Ivoire. *Bull. IFAN,* Série A, 30(3): 1038-1151.

Hairston, N. G. (1989) *Ecological Experiments.* Cambridge University Press, New York, NY.

Helfer, J. R. (1963, 1972) *How to Know the Grasshoppers, Cockroaches and Their Allies.* 2nd edition. Wm. C. Brown Company Publishers, Dubuque, IA.

Hellweg, M. R. (2009) *Raising Live Foods.* T.F.H. Publications Inc., Neptune City, NJ.

Honsa, V. (2003) *Acromantis* sp. (Malaysia, Landkani). *Sklipkan Journal for Invertebrates Keepers and Breeders* (Czech Republic) 8(2).

Howarth, F. G., and Mull, W. P. (1992) *Hawaiian Insects and Their Kin.* University of Hawaii Press, Honolulu, HI.

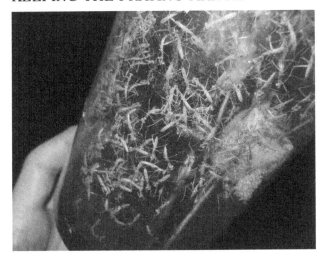

Hatchlings from one *Tenodera sinensis* ootheca cover the sides and lid of a 32oz. deli cup (318 nymphs). A much, much larger growout cage is needed. Nymphs should be able to spread out enough to have a half inch space between each, if not more.

Hurd, L. (1985) Ecological considerations of mantids as biocontrol agents. *Antenna* 9(1): 19-22.

—. (1999) Ecology and mating behavior. In: Prete, F. R., Wells, H., Wells, P. H., and L. E. Hurd. (1999) *The Praying Mantids*. Johns Hopkins University Press, Baltimore and London. pp. 43-60.

Hurd, L. E., and R. M. Eisenberg. (1984) Experimental density manipulations of the predator *Tenodera sinensis* (Orthoptera: Mantidae) in an old-field community. I and II. The influence of mantids on arthropod community structure. *J. Anim. Ecol.* 53: 269-281, 53: 955-967.

—. (1990) Arthropod community responses to manipulation of a bitrophic predator guild. *Ecology* 76: 2107-14.

Hutchins, R. E. (1966) *Insects*. Prentice Hall, Englewood Cliffs, NJ.

Imes, R. (1992) *The Practical Entomologist: An Introductory Guide to Observing and Understanding the World of Insects*. Simon and Schuster, New York, NY.

Inward, D., Beccaloni, G., and P. Eggleton. (2007). Death of an order: a comprehensive molecular phylogenetic study confirms that termites are eusocial cockroaches. *Biology Letters* 3(3): 331-335.

Jiang, W. (2006) Master of mimicry: *Phyllocrania paradoxa*. *Invertebrates-Magazine* 5(3).

Kaltenbach, A. P. (1963) Kritische untersuchungen zur systematik, biologie und verbreitung der europäischen fangheuschrecken (Dictyoptera-Mantidae). *Zool. Jahrb. Syst.* 90: 521-598.

Kaltenbach, A. P. (1996) Unterlagen für eine monographie der Mantodea des südlichen Afrika: 1.– Artenbestand, geographische verbreitung und ausbreitungsgrenzen (Insecta: Mantodea). *Ann. Nathist. Mus. Wien.* 98(B) 193-346.

Kral, K. (1998) Side-to-side head movements to obtain motion depth cues: A short review of research on the praying mantis. *Behav. Proces.* 43: 71-77.

Kral, K., and M. Poteser. (1997) Motion parallax as a source of distance information in locusts and mantids. *Journ. Ins. Behav.*, 10(1): 145-163.

Kurtz, Carey (2013) The Indonesian double shield mantis: breeding a very aggressive female mantis. *Invertebrates-Magazine* 12(2): 5-7.

Larsen, T. (2001) The Chinese mantis—*Tenodera aridifolia sinensis*. *Invertebrate* #00 (Fall).

—. (2002) The wandering violin: spotlight on *Gongylus gongyloides*. *Invertebrate* #02 (Spring): 4-9.

—. (2007) *Knaelere*. Narayana Press, Denmark.

Lasebny, A., and O. McMonigle. (2001) *Praying Mantids: Keeping Aliens*. 1st Edition. Elytra and Antenna, Brunswick, OH.

Ladau, J. (2003) Prey capture in a mantid (*Gonatista grisea*): does geotropy promote success? *Can. J. Zool.* 81: 354-365.

Linnaeus, C. (1758) *Systema Naturae*. Holmiae, Laur. Salvius, Stockholm.

Mangels, J. (2012) Bugs: Cleveland to be hub of praying mantis study. (Cleveland, OH) *The Plain Dealer*. (July 15).

Maxwell, M. R., Barry, K. L., and P. M. Johns. (2010a) Examinations of female pheromone use in two praying mantids, *Stagmomantis limbata* and *Tenodera aridifolia sinensis* (Mantodea: Mantidae). *Ann. Entomol. Soc. Am.* 103(1): 120-127.

Maxwell, M. R., Gallego, K. M., and K. L. Barry. (2010b) Effects of female feeding regime in a sexually cannibalistic mantid: fecundity, cannibalism, and male response in *Stagmomantis limbata* (Mantodea). *Ecol. Entomol.* 35: 775-787.

McMonigle, O. (2001) A deadly flower: the orchid mantis. *Reptile and Amphibian Hobbyist* 7(3): 60-64.

—. (2004a) The Malaysian deadleaf mantis. *The Keeper: A Monthly Paper for Keepers of Reptiles, Amphibians and Invertebrates.* (July): 1, 14.

—. (2004b) Featured invertebrates: Brunner's mantis *Brunneria borealis*, Arizona unicorn mantis *Pseudovates arizonae*. *Invertebrates-Magazine* 4(1): 5-6.

—. (2006) *Giant Tarantulas: The Enthusiast's Handbook*. Elytra and Antenna, Brunswick, OH.

—. (2011) *Invertebrates for Exhibition: Insects, Arachnids, and Other Invertebrates Suitable for Display in Classrooms, Museums, and Insect Zoos*. Coachwhip Publications, Landisville, PA.

—. (2012a) Teneral mating as a standard function of copulation in the Dictyoptera. *Invertebrates-Magazine* 12(1): 2-4.

—. (2012b) A pretty mantis, *Sibylla pretiosa* Stal 1856. *Invertebrates-Magazine* 12(1): 9-11.

McMonigle, O., and Y. Saw. (2007) The last unicorn (mantis). *Invertebrates-Magazine* 6(4).

Milledge, G. A. (1997) Revision of the tribe Archimantini (Mantodea: Mantidae: Mantinae). *Memoirs of the Museum of Victoria* 65(1): 1-63.

Moran, M., Rooney, T., and L. Hurd. (1996) Top down cascade from a bitrophic predator in an old field community. *Ecology* 77: 2219-27.

Otte, D., Spearman, L., and M. B. D. Stiewe. (2012) *Mantodea Species File Online*. Version 1.0/4.1. [10,1,2012]. http://Mantodea.SpeciesFile.org.

Stagmomantis limbata.

Swan, L., and C. Papp. (1972) *The Common Insects of North America*. Harper and Row Publishers, New York, NY.

Perez, B. (2005) Calling behaviour in the female praying mantis, *Hierodula patellifera*. *Physiological Entomology* 30: 42-47.

Poteser, M., and K. Kral. (1995) Visual distance discrimination between stationary targets in praying mantis: an index of the use of motion parallax. *Journ. Exp. Biol.* 198: 2127-2137

Powell, J., and C. Hogue. (1979) *California Insects*. University of California Press, Berkeley and Los Angeles, CA.

Preston-Mafham, K. (1990 and 1992) *Grasshoppers and Mantids of the World*. Blandford Pub., London.

Preston-Mafham, K., and R. Preston-Mafham. (2000) *Bug and Insects*. PRC Publishing Ltd., London.

—. (2005) *Encyclopedia of Insects and Spiders*. Thunder Bay Press, San Diego, CA.

Prete, F. R., and M. M. Wolfe. (1992) Religious supplicant, seductive cannibal, or reflex machine? In search of the praying mantis. *Journ. Hist. Biol.* 25(1): 91-136.

Prete, F. R., Wells, H., Wells, P. H., and L. E. Hurd. (1999) *The Praying Mantids*. Johns Hopkins University Press, Baltimore, MD.

Prokop, P., and R. Vaclav. (2005) Males respond to risk of sperm competition in the sexually cannibalistic praying mantis, *Mantis religiosa*. *Ethology* 111: 836-848.

Ramsay, G. W. (1984) *Miomantis caffra* a new mantid record (Mantodea: Mantidae) from New Zealand. *New Zealand Entomologist* 8: 102-104.

Roeder, K. D. (1935) An experimental analysis of the sexual behavior of the praying mantis (*Mantis religiosa* L.). *Biol. Bull.* 69: 203-220.

Ross, E. S. (1984) Mantids the praying predators. *National Geographic.* 165(2): 268-280.

Roy, R. (1977) Mises au point sur le genre *Heterochaeta* Westwood (Mantodea). *Bull. IFAN*, Série A, 38(1): 69-111.

—. (2004) Réarrangements critiques dans la famille des Empusidae et relations phylogénétiques (Dictyoptera, Mantodea). *Revue française d'Entomologie (N. S.)* 26(1): 1-18.

Schwarz, C. J. (2003) *Diets and Habitat Preferences of Neotropical Praying Mantids (Dictyoptera: Mantodea Burm. 1838)*. Diploma thesis, 114 p.

—. (2004) *Popa spurca spurca* Stål, 1856 – eine bemerkenswerte Gottesanbeterin aus Westafrika (Mantodea: Mantidae: Vatinae). *Arthropoda* 12(2): 12-20.

Schwarz, C., Mehl, J., and J. Sommerhalder. (2009) *Idolomantis diabolica* (Saussure, 1869) The devil's flower. Part one and two. *Invertebrates-Magazine.* 8(2, 3).

Sharp, D. (1899) The modification and attitude of *Idolum diabolicum*, a mantis of the kind called "floral simulators". *Proc. Cambr. Phil. Soc.* 10: 175-180.

Skoupy, V. (2002) *Deroplatys desiccata*. *Sklipkan Journal for Invertebrates Keepers and Breeders* (Czech Republic) 9(2).

Stanek, V. J. (1969) *The Pictorial Encyclopedia of Insects*. Hamlyn Publishing, London.

Stefferud, A. (1952) *Insects: The Yearbook of Agriculture*. United States Department of Agriculture, Washington D. C.

Stagmomantis gracilipes. © Peter Clausen

Svenson, G., and M. Whiting. (2009) Reconstructing the origins of praying mantises (Dictyoptera, Mantodea): the roles of Gondwanan vicariance and morphological convergence. *Cladistics* 25(2009): 468–514.

Symes, G. (1994) Demi-johns as breeding cylinders for mantids, etc. *The Bulletin of the Amateur Entomologists' Society* 53(392): 36.

Thomann, C. H. (2002) Central-southeastern U.S. mantid *Stagmomantis carolina* (Johansson 1763). *Invertebrate* #04.

Tinnesen, K. (2005) Care and breeding for the mantis *Ceratomantis sausurii*. *Invertebrates-Magazine* 4(2).

Tomasinelli, F. (2002) The banded flower mantid: biology and captive breeding of *Theopropus elegans*. *Invertebrate* #04.

—. (2003) Angel of death, the orchid mantis mimicry, behavior and reproduction of *Hymenopus coronatus*. *Invertebrates-Magazine* 2(2).

Triblehorn, J. D., and D. D. Yager. (2001) Broad vs. narrow auditory tuning and corresponding bat-evasive behaviour in praying mantids. *Journal of Zoology* 254: 27–40.

—. (2005) Timing of praying mantis evasive responses during simulated bat attack sequences. *J. Exp. Biol.* 208: 1867-1876.

Watanabe, H., and E. Yano. (2009) Behavioral response of mantid *Hierodula patellifera* to wind as an antipredator strategy. *Ann. Entomol. Soc. Am.* 102(3): 517-522.

—. Behavioral response of male mantid *Tenodera aridifolia* (Mantodea: Mantidae) to windy conditions as a female approach strategy. *Entomol. Science*, in press.

—. Behavioral response of mantid *Tenodera aridifolia* (Mantodea: Mantidae) to windy conditions as a cryptic approach strategy for approaching prey. *Entomol. Science*, in press.

Wieland, F. (2008) The genus *Metallyticus* reviewed (Insecta: Mantodea). *Species, Phylogeny and Evolution* 1(3): 147-170.

Willis, R. (1999) *Your First Praying Mantis*. Kingdom Books, Havant, U.K.

Yager, D. D. (1996) Serially homologous ears perform frequency range fractionation in the praying mantis, *Creobroter* (Mantodea, Hymenopodidae). *J. Comp. Physiol. A.* 178(4): 463-75.

—. (1999) Hearing. In: Prete, F. R., Wells, H., Wells, P. H., and L. E. Hurd (eds.): *The Praying Mantids*. Johns Hopkins University Press, Baltimore and London. pp. 93-113.

Yager, D. D., and R. R. Hoy. (1986) The cyclopean ear: a new sense for the praying mantis. *Science* 231(4739): 727-729.

—. (1987) The midline metathoracic ear of the praying mantis, *Mantis religiosa*. *Cell and Tissue Res.* 250: 531-541.

Yager, D. D., and M. L. May. (1990) Ultrasound-triggered, flight-gated evasive maneuvers in the praying mantis *Parasphendale agrionina* (Gerst.). II. Tethered flight. *Journ. Exp. Biol.* 152: 41-58.

Yager, D. D., and G. J. Svenson. (2008) Patterns of praying mantis auditory system evolution based on morphological, molecular, neurophysiological and behavioural data. *Biol. Journ. Linn. Soc.* (94): 541-568.

Coachwhip Publications
CoachwhipBooks.com

The Ultimate Guide to Breeding Beetles
Orin McMonigle

ISBN 978-1616461324

Orin McMonigle provides detailed husbandry and breeding guides for a wide range of beetle species, from the popular rhinoceros and stag beetles to darkling, diving, and dung beetles. This book is the result of years of experience and experimentation, with unprecedented details in caging, feeding, and environmental requirements for all stages of the beetles' lives. The breeding guides offer the best chance to form healthy ongoing colonies of these incredible creatures. This is the ultimate beetle book for hobbyists, breeders, nature museums, and insect zoos. Welcome to the world of beetles!

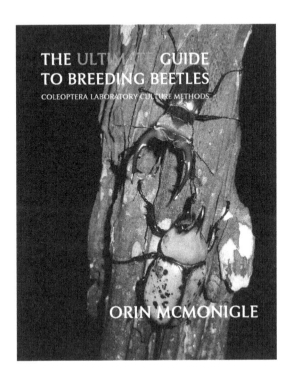

Millipeds in Captivity
Orin McMonigle

ISBN 978-1616461430

Orin McMonigle, with contributions by the late Dr. Richard L. Hoffman, assembles the definitive resource guide with reproductive and developmental data for those spectacular terrestrial arthropods, the millipeds (or millipedes). Invertebrate hobbyists can successfully culture a number of colorful and gigantic diplopods by following specific methodologies outlined in this book. From the world's largest African giant millipeds to the most astoundingly colorful members of the Orders Polydesmida and Spirobolida, there are plenty of species to attract the beginning enthusiast or to challenge the advanced keeper.

Lightning Source UK Ltd.
Milton Keynes UK
UKOW07n2118050515

250908UK00007B/100/P